A Traveler's Guide
to the
Geology
of Egypt

To My Dear Frind Mark Sims

Mahmoud Khodier

6/13/2014.

A Traveler's Guide to the Geology of Egypt

Bonnie M. Sampsell

The American University in Cairo Press
Cairo • New York

Dar el Kutub No. 19213/02
ISBN 978 977 424 785 9

   3 4 5 6 7 8 9 10 11 12   12 11 10 09 08 07

Designed by Andrea El-Akshar/AUC Press Design Center
Printed in Egypt

Contents

Tables and Illustrations

Tables

Figures

Tables and Illustrations

Color Photographs (between pages 164 and 165)

Satellite view of Lake Nasser, Toshka lakes, and the southern Nile Valley
The Grand Canyon in Arizona
Pediplain landscape along Lake Nasser
Granite boulders along the Nile near Aswan
Coarse-grained red granite from Aswan
Holes prepared for wedges to split a granite block

Dolerite pounder used to shape granite
Hole for a dovetail cramp in a cracked statue of Ramesses II
Graffiti on a block of easily carved Nubian sandstone
Hatshepsut's Mortuary Temple, and the contact between Esna Shale and
    Theban Limestone
Unfinished quartzite statue found in the quarry of Gebel el-Ahmar, Cairo
The northern colossal quartzite statue of Amenhotep III
Salt efflorescence on an inscribed sandstone wall in Karnak Temple
Spalled ceiling, cracked pillar, and flood debris in KV 10
Inscribed travertine façade, barque chapel of Tuthmosis IV
Reconstructed Red Chapel of Hatshepsut
Inscribed blocks of the Red Chapel
A shadoof in use
Pigeons in a dovecote, source of food and fertilizer
A donkey-powered waterwheel lifts water from a well
Satellite view of the northern Nile Valley, the Fayum, and the Delta
Fine-grained limestone, probably from a quarry at Tura
Limestone filled with fossil nummulites on the Giza Plateau
Spalling of limestone
Basalt paving block in Khufu's Mortuary Temple
Fine-grained basalt
Stumps left after the removal of limestone blocks near Khafre's Pyramid
Differential erosion on the Sphinx
The granite stela erected in front of the Sphinx by Tuthmosis IV
Erosion on Old Kingdom tombs at Giza
Siwa Oasis viewed from the ruined village of Aghurmi
Sculpted limestone rock in the White Desert near Farafra Oasis
St. Anthony's Monastery and the escarpment of the Southern Galala Plateau
Gezirat Fara'un (Pharaoh's Island) in the Gulf of Aqaba
Sunrise on the granite peaks in southern Sinai
Satellite view of the Sinai Peninsula and the Red Sea

Introduction
Why Study Geology?

A s a traveler in Egypt, you may have a keen interest in the country's culture and history. You may plan to visit monuments and ruins dating to
the pharaonic, Classical, Christian, and Islamic eras, to visit museums to see
the artifacts found in these sites, and to listen to Egyptologists discuss religion, art, and architecture.

Alternatively, you may have come to Egypt to relax—to enjoy the
scenery, the wonderful climate, to hike, to fish, or to dive in the Red Sea.
Whatever your objective, you will be surrounded by a landscape at once
ancient and modern—a country in which 4,500-year-old monuments provide
a backdrop for contemporary villages going about their everyday business.

Many travelers find that their understanding of a country is augmented and
the pleasure of their trip enhanced if they also appreciate the environment—
the flora, fauna, and landforms that distinguish it. Ancient Egypt, like every
civilization, was influenced by the physical setting in which it developed. The
Greek historian, Herodotus, visiting in about 450 BCE, said it best when he
wrote: "Egypt is a gift of the Nile." Later societies, including that of today,
have also been dependent on the Nile's bounty and constrained by other geological factors in the country.

In fact, four features dominate the landscape in Egypt: the River Nile, the
rocky cliffs, the rugged mountains, and the immense deserts. The importance
of the Nile for Egyptian society as the source of its fertile land and its major
transportation artery is obvious to all students of Egyptian history, but few
people know anything about the history of the river itself and how it came to
assume its current form. The river's past holds important clues about its future.

At first glance, the rocks of the cliffs bordering the river may seem to have
little bearing on the development of civilization. Comparisons between Egypt

and Mesopotamia, the other ancient river-based civilization, demonstrate that the presence of building-stone in Egypt (but not in Mesopotamia) influenced the style of Egypt's architecture and art, stimulated the development of its technology and transport, and enabled it to leave a permanent record for later generations. Why is there so much stone in Egypt, and why are the different kinds located where they are? In this book you will find out.

Egypt's mountains and desert served as barriers against invading foreigners. At the same time they also permitted some habitation and served as a source of valuable rocks, minerals, and metals. Could these deserts now provide living space for Egypt's growing population?

To understand why a country has the particular landscapes and landforms it does—whether mountains or plains, river valleys or deserts, and the extent of its mineral deposits—requires an examination of its history stretching back eons before the first human settlement. Geology can supply that historical perspective. But to many people geology seems a daunting topic, full of specialized vocabulary and unfamiliar concepts. This book is intended for the general traveler, so I have avoided technical jargon wherever possible. The two biggest groups of terms that cannot be avoided are those for the different intervals of the Geological Time Scale and for the names of rocks and minerals. Several tables present these terms in an organized fashion and should be consulted frequently. Terms printed in **bold type** in the text are defined in the Glossary.

The book begins with a chapter containing background material about important geological processes and about rock composition and formation. These ideas are then applied to understand how the unusual Egyptian landscape evolved. Then there is a chapter about the River Nile—its history, its recent past, and its effects on Egyptian civilization. The balance of the book contains chapters dealing with the geology of individual sites or regions that you are likely to visit.

You will learn that the disciplines of geology and archaeology can interact in a productive way. The techniques of geology, such as rock dating and the formation of hypotheses about ancient environments, can assist the archaeologist in creating a picture of an ancient society. Geologists are now routinely included in the multidisciplinary teams on archaeological expeditions.

This book is designed to serve as an introduction to the geology of Egypt for the non-specialist, but I still felt it was important to include the latest information and scientific hypotheses. At the same time, much remains to be explained, and many details that are the subject of continuing study and debate among geologists have not been included. Most examples of the impact of rocks and water on Egyptian civilization are taken from the dynastic period, since that is what interests most tourists. Examples from other periods are used where they are appropriate and particularly instructive. In so brief a work it is not possible to give any kind of overview of the extensive archaeological remains at places such as Giza, Saqqara, and Luxor, and a certain familiarity with the main sights must be assumed. The reader is urged to consult any of the excellent books about these sites that are available.

Metric units have been used throughout the book since this is the system used by geologists and archaeologists. For those more comfortable with non-metric units, here are some rough equivalents: a meter (m) is a little more than a yard, a kilometer (km) is a little more than half a mile, while one inch equals 2.5 centimeters (cm) or 25 millimeters (mm). More information is given in the Glossary under **metric units**.

The spellings of place names are taken from the *Geological Map of Egypt*, published in 1986–87. In a few cases, I have preferred an alternative spelling drawn from Rushdi Said's *Geology of Egypt, 1962*. You may encounter variations for some of these spellings, but the meaning is generally clear.

Dates are given as BCE ('Before the Common Era,' equivalent to BC) and CE ('Common Era,' equivalent to AD).

Important Geological Concepts

Geologists divide the immense span of Earth's history into discrete units of time. The largest or longest units are called **Eons**. Table 1.1 lists two Eons: the Phanerozoic and the Precambrian. The term Phanerozoic is derived from the Greek words for 'appearing' and 'animals.' The Phanerozoic Eon began around 590 million years ago when organisms with mineralized skeletons began to be preserved as fossils. The Phanerozoic Eon is divided into three subdivisions called **Eras**: the Paleozoic or 'old animals,' the Mesozoic or 'middle animals,' and the Cenozoic or 'recent animals.' Eras can be further divided into **Periods**, which are divided into **Epochs**. The term Precambrian is an informal designation for the entire time before the Cambrian Period. It began approximately 4,600 million years ago when the Earth's crust cooled.

Rocks that are formed during a particular time interval can be characterized by one or more of the following criteria: their absolute age, their position relative to rocks above and below them, their composition, and their fossil inclusions. Changes in one or more of these criteria are used to define the boundaries between time units. Geologists continue to adjust the boundaries as they find new evidence and in an effort to achieve a system with worldwide application. As a result, charts from different sources sometimes vary slightly in the dates they assign to the beginnings of various time units. For simplicity, I have included in Table 1.1 only those intervals relevant to material discussed in this book; many other subdivisions have been defined and named.

1

Table 1.1: The Geological Time Scale

Eon	Era	Period	Epoch	Beginning (years ago)
Phanerozoic	Cenozoic	Quaternary	Holocene	10,000
			Pleistocene	2 million
		Tertiary	Pliocene	5.1 million
			Miocene	24.6 million
			Oligocene	38 million
			Eocene	55 million
			Paleocene	65 million
	Mesozoic	Cretaceous		144 million
		Jurassic		213 million
		Triassic		248 million
	Paleozoic	Permian		286 million
		Carboniferous		360 million
		Devonian		408 million
		Silurian		438 million
		Ordovician		505 million
		Cambrian		590 million
Precambrian				c. 4,600 million

Based on Allaby and Allaby, 1991.

Plate Tectonics and Continental Drift

Beginning around 1960, geologists began to develop a theory of **plate tectonics** that explained and unified many processes and events within their discipline. One of the main components of this theory is the idea that the solid outer surface of the Earth, called the **lithosphere**, is divided into a number of contiguous sections, called 'plates' since they are thin in comparison to their lateral dimensions (Fig. 1.1). Most plates contain a landmass as well as parts of its surrounding ocean basins, but some plates and microplates may be entirely oceanic.

The rigid plates are moving in different directions over the underlying semi-molten **mantle** at average rates of a few centimeters a year. The motive force causing the plates' motion is still not well understood, but it appears to involve thermal convection currents in the mantle that lead to an upwelling of hot molten rock at certain locations. The consequence of this plate motion is

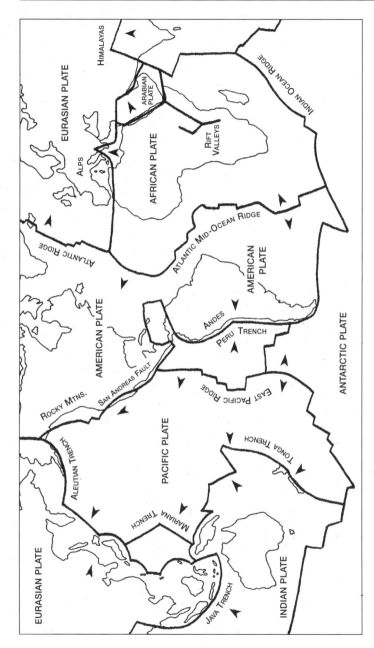

Fig. 1.1: The major lithospheric plates showing the current positions of their boundaries and directions of movement as indicated by arrowheads. Diverging zones between two plates are characterized by mid-ocean ridges. Convergence zones where two plates are colliding are marked by deep ocean trenches and/or mountain ranges.

the slow but ongoing rearrangement of the Earth's surface with the continual formation of new land and oceanic **crust** at some points while oceanic crust in other locations is being destroyed. The sizes of the plates are not constant but change over time.

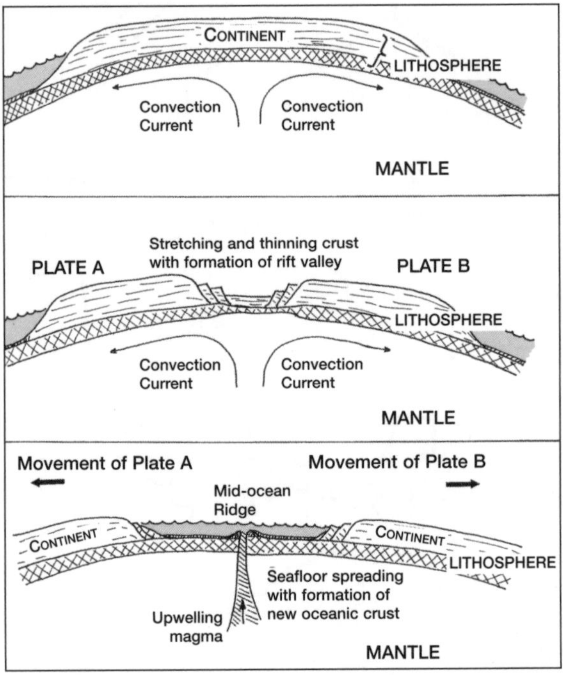

Fig. 1.2: The rifting of a lithospheric plate within a continent to produce two plates and eventual seafloor spreading.

If upwelling occurs under a plate, the continental crust above may form a dome or may simply thin and stretch (Fig. 1.2). This stretching can crack the crust in several places and allow blocks of crust to sink, forming a **rift valley**. Molten **magma** may seep through these cracks onto the land surface. Eventually the original plate may split into two receding plates, with an expanding ocean basin between them. Molten magma rising through cracks in oceanic crust will harden and form new oceanic crust along a distinctive mid-ocean ridge.

If two plates move toward one another (Fig. 1.3), they will collide and weld their landmasses together; meanwhile the intervening oceanic crust is destroyed as one plate slides beneath (is **subducted** by) the other one. The oceanic crust on the subducted plate melts as it descends into the hot man-

4

tle, and the molten material rises to join the continental crust above. **Sediments** that have washed off the land and collected along the continental margin may be scraped up and added to the welded continent. Such collision zones are often marked by mountain ranges built from volcanoes and uplifted continental crust. Deep ocean trenches are also characteristic of some collision zones.

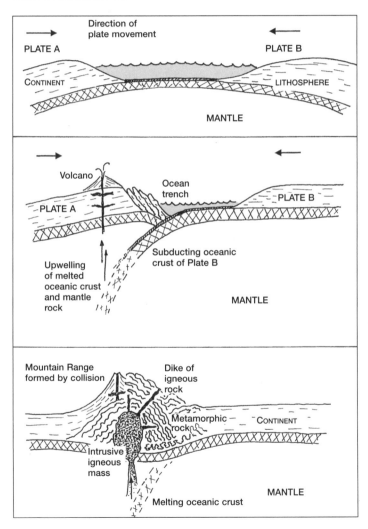

Fig. 1.3: The convergence or collision of two lithospheric plates with the formation of an ocean trench (offshore) and a mountain range (onshore).

Much of the continental landmass of Africa was formed during the Precambrian. Three small plates with pieces of ancient continental crust, termed **cratons**, collided and were welded together. One of these cratons (called the Nile craton) extended into southwest Egypt. This area contains the oldest rocks in the country, dated at more than 2,000 million years old. The area of Egypt and other parts of Africa were then enlarged by the addition of more material at the margins of this African Plate, a process that continued through the Phanerozoic Eon and up to the present.

As the continents pursue their random paths around the globe, they have occasionally coalesced into a gigantic continent containing all the earth's land-masses. The last time this occurred was around 300 million years ago, forming the continent called **Pangaea**. The collisions that brought Africa into this super-continent, where it occupied a central position, added material to its margins and severely deformed the rocks already present. Around 200 million years ago Pangaea began to experience rifting and to break apart. Little by little Pangaea was separated into the continents we know today. Since then they have contin-ued to move along separate paths.

Geologists have been able to project the locations of various continents backward to see where they were located at various times in the past. In its wan-derings, Africa has been at the South Pole and at the Equator. There is evidence that around 480 million years ago the northeast corner of Africa (that is, Egypt) lay at 60° south latitude; today it lies between 22° and 32° north of the Equator. Such a northward shift in Africa, as well as the significant rearrangement of the other continents, must have had a very profound effect on the climate of Egypt.

After the complete breakup of Pangaea, the African Plate began to move closer to the Eurasian Plate. The movement of the African Plate was general-ly to the northeast, but during the last 10 million years that changed to a north-westerly course relative to Europe. This direction will become important when we consider the collision between these two plates that is now under-way. The long, narrow ocean separating the landmasses of Europe and Africa was called the Tethys Sea. As the continents approached each other the sea's width shrank through subduction of oceanic crust. The Mediterranean Sea is the remains of this contracting ocean, and its several deep basins are all that remain of the former seabed.

Today the African Plate is still being subducted beneath the Eurasian Plate. The well-known earthquakes of the Mediterranean region are the consequences of these **tectonic** movements. The collision of the two plates has also been responsible for the formation of mountain ranges including the Atlas, Alps, Apennines, Pyrenees, and Carpathians, while the melted subducted material fuels the volcanism of Mounts Etna, Vesuvius, and Santorini among others.

Kinds of Rocks

Rocks are the basic building blocks of the earth's crust. Geologists recognize hundreds of kinds of rocks, distinguishing them by their method of formation, textures, and chemical composition. By studying the rocks of an area, a geologist can reconstruct the history of that place. Familiarity with just a few rock types will be sufficient for our purposes. Based on their method of formation, we classify rocks as **igneous**, **sedimentary**, or **metamorphic**.

Igneous rocks are produced when molten rock, called magma, cools and solidifies. Magma forms deep within the earth, frequently in zones where two plates are colliding and one is being subducted below the other. The heat at great depths melts the subducted crust. This magma is less dense than solid rock, and it tends to rise back toward the surface, where it cools and solidifies. Geologists have identified dozens of kinds of igneous rocks that differ in their mineral content and crystalline texture; examples of some common kinds are listed in Table 1.2, where they are also categorized as either **intrusive** or **extrusive** depending on whether their solidification takes place underground or above ground.

Table 1.2: Igneous rocks

Igneous rock type	Forms at	Rate of cooling	Crystalline nature	Examples
extrusive	surface of Earth	rapid	small crystals	rhyolite andesite basalt
intrusive	shallow depths	moderate	medium crystals	dolerite
			small + medium/large	porphyry
intrusive	great depths	slow	large crystals	granite granodiorite diorite gabbro anorthosite

Intrusive igneous rocks form when the magma cools below ground; there the rate of cooling is slow enough to permit large crystals of different minerals to develop within each rock. Although these rocks originate below ground, they can later be uplifted, which will encourage erosion that will expose them at the surface. Tectonic uplift can occur by additional magma upwelling or by elevation during a mountain building event.

Extrusive (or **volcanic**) igneous rocks are formed when magma, called **lava**, reaches the surface of the earth through **faults** or channels that feed volcanoes. Depending on its chemical composition, lava may be thick and viscous or quite liquid. Thin lava will easily escape through cracks in the earth's crust; it can flow for great distances, forming sheets.

The formation of some **sedimentary** rock types also involves a recycling of old rocks into new ones, but the old rocks are not melted; they are simply broken up into fragments of various sizes. These fragments are transported by wind or water until they are deposited in streambeds, river channels, floodplains, landlocked basins, coastal lagoons, or sea floors. The fragments of older rocks become new rock when they are bound or cemented together by chemicals precipitating from the seawater or from ground water percolating through the deposits. Common rock cements include calcium carbonate, silica (silicon dioxide), and iron oxide. Different sedimentary rocks are characterized by their kind and size of fragments and their type of cement. Table 1.3 gives a list of sedimentary rocks formed of rock fragments, along with the type of particles composing them and the size of these particles.

Table 1.3: Sedimentary rocks formed from rock fragments

Sedimentary rock type	Particle name	Smallest-sized grain
Conglomerate	Gravel	256 mm (boulder) 64 mm (cobble) 4 mm (pebble) 2 mm (granule)
Sandstone	Sand	0.063 mm
Siltstone	Silt	0.004 mm
Shale	Silt Clay	0.004 mm less than 0.004 mm

A sedimentary rock's properties are a function of its fragments and its cement. For example, sandstones vary in the kind and quantity of cement binding their sand grains. As a result they vary in their hardness, which affects how well they perform as building stone. In some sandstones the cement is weak and the individual grains separate readily from their neighbors; in others the cement may actually be tougher than the material it holds together, so that when the rock breaks, it breaks across the grains.

One of the most important sedimentary rocks in Egypt is limestone. This rock is not defined by the size of its particles but by its chemical composition: both its grains and cement are composed of the compound calcium carbonate. **Calcite** and **aragonite** are common minerals containing this compound. Limestone forms in a variety of marine environments. Many marine organisms ranging in size from giant clams and other molluscs down to tiny protozoa extract calcium carbonate from seawater and use it to form external shells or internal skeletons. When these organisms die, their shells or broken shell fragments sink to the bottom, where they may be **lithified** by calcium carbonate precipitating from saturated seawater or produced by other limestone-secreting organisms. The texture of limestones can vary from very coarse (when shell fragments are large) to very fine (when the grains are tiny or sand-sized). Limestone is generally considered a relatively soft rock (compared to granite, for example), but this is a function of the amount of calcium carbonate cement; some poorly cemented limestones are very friable or **chalky**, while other types are much harder. Limestone may contain considerable amounts of sand or clay (see **marl**). Its purity, texture, and hardness will determine a limestone's suitability for use as a building stone, for statuary, or for other purposes.

Several other kinds of sedimentary rock consist entirely of minerals precipitated from mineral-rich seawater or evaporating pools; such rocks are very fine-textured. **Flint** (or chert) is found within layers of some limestone in the form of seams or nodules. Such nodules form when silica replaces the calcium carbonate. **Gypsum** and halite (rock salt) are usually formed by evaporation; they consist of calcium sulfate and sodium chloride respectively. Travertine forms when water saturated with calcium carbonate evaporates around a hot spring or in a cave to form stalactites and stalagmites.

Layers of sediment are usually horizontal when they are deposited; such a layer is called a **stratum** (plural **strata**) or a bed. Younger layers accumulate on top of older layers. These two facts have been referred to as the 'law of horizontality' and the 'law of superposition' of strata and are keys to an analysis of rock origins and dating. After sediments have become cemented into stone, tectonic events such as plate collisions may change their orientation—uplifting and tilting or even overturning blocks composed of many layers. Geologists learn to recognize these disturbances and use them as evidence to reconstruct past events. Younger rocks can also underlie older ones when magma rises into older rocks: the igneous rocks will be younger than their surroundings. The study of rock strata is an important branch of geology called **stratigraphy**.

If sediment continues to be deposited at a location and **lithification** (cementing) continues, the depth of the sedimentary rock will increase with time. As conditions change the kind of rock formed at a particular location will also change. For example, the kind and quantity of fragments deposited by a river into an ocean depend on the conditions in its **drainage basin**; these conditions include the kinds of rocks exposed, the climatic conditions, the volume of water flowing, the slope of the river, and other factors. When the river meets the sea, coarse sediments, that is large particles, will be deposited close to shore, while fine particles will be carried out farther before they finally sink to the bottom. Deep basins will permit sediments to be deposited for longer periods, building up thicker layers of rock. Shallow basins will fill quickly so only thin beds result.

Once they are formed, both igneous and sedimentary rocks are subject to additional processes that can change their chemical and/or physical nature. We have seen that the earth's crust is subject to many tectonic forces: uplift, folding, rifting, and collisions. These forces cause cracks in the brittle crust. If the pieces of crust on either side of the crack move—either up and down or slide past each other—we call the crack a **fault**. If the pieces do not move, we call the crack a fracture. Fractures can also form in igneous rocks as a way of relieving stresses when they cool or when pressure on them is reduced. Sedimentary rocks develop fractures as the sediments harden and shrink.

If a rock is subjected to heat and pressure its nature can be altered so much that we place it in the third major rock class: **metamorphic**. Each original rock type gives rise to a corresponding metamorphic form. Generally the conditions required to effect this transformation can only be found deep underground, where plates are colliding or when still molten magma comes into contact with surrounding rocks. The latter situation is referred to as 'contact metamorphism.' Note that while rocks are heated during metamorphic processes they are not melted.

Table 1.4: Some examples of metamorphic rocks

Original rock type	Rock type after metamorphosis
Granite, other igneous rocks	Gneiss
Sandstone	Quartzite
Shale	Slate
Limestone	Marble

Weathering, Erosion, Transport, and Deposition

In contrast to metamorphic changes, which require heat and pressures not normally found on the earth's surface, the processes of rock **weathering** and **erosion** occur at or near the earth's surface. These processes can be accentuated by previous cracking of the rocks. **Weathering** is the term that refers to the breakdown or decomposition of rocks on or near the surface of the earth by either chemical or physical means. The term reflects the fact that atmospheric agents are usually involved. Chemical weathering involves a chemical reaction between the minerals of the rock and water along with certain atmospheric gases such as oxygen and carbon dioxide. Organic acids produced by decaying plants can also produce these reactions. These chemical reactions occur most rapidly in areas of high temperature and humidity. The products of the chemical reactions are either smaller molecules or more soluble minerals than the original, permitting their removal by wind or water.

Physical weathering also breaks rocks down into finer fragments, but without any change in their chemical composition. One kind of physical weathering described in older geology books is thermal stress. It proposed that if rocks go through a daily temperature range, they expand and contract, which causes stress. Rocks are poor conductors of heat, so sun-warmed rocks heat and expand on the surface more than on the inside, fracturing off layers, especially where the rock is already weakened by chemical processes. This form of weathering was thought to be very important in hot desert regions where the air temperatures show a great daily range. While this process seems eminently reasonable and many geologists working in the desert report hearing cooling rocks 'popping,' newer books report that the process cannot be duplicated in the laboratory unless moisture is present. Nonetheless, fractured rocks are very common in hot deserts, whatever the precise mechanism.

When moisture freezes in rock crevices, it expands and causes cracking; thawing allows the moisture to creep deeper into the cracks. In only a few parts of Egypt, such as the Sinai Mountains, does such frost weathering occur. The most common type of weathering in dry, hot climates occurs when water carrying dissolved salts seeps from the soil up into rocks through capillary action. When the water evaporates from the surface of the rock, large salt crystals form. When the crystals form on the surface of the rock, the process is called efflorescence, while if the crystals form just below the surface, it is called subflorescence. This is a more serious situation, since it can cause the rock surface to split and **spall** or detach in flakes. We will encounter these processes frequently as we study particular locations.

Once a rock has weathered, the weathered layer can be removed by wind (a process known as **deflation**), running water or wave action, or moving ice (glaciers). Such weathering and removal can be referred as **erosion**. Once rock fragments are transported, we will refer to them as **sediments**, since eventually they will come to rest—either in an inland basin, in a river channel, or in the ocean.

There is a complex relationship between annual precipitation and the volume of sediment produced in an area. At the lowest levels of precipitation, there is little chemical weathering or runoff, although the absence of rainfall prevents growth of vegetation and means that the rocks are exposed. Increasing precipitation increases both weathering and runoff, thereby increasing sediments. At still higher levels of precipitation, however, grasses grow and protect the ground's surface against erosion by either wind or water. At the highest levels of precipitation, the grasses are replaced by trees that shade out the vegetative ground cover, thus permitting more runoff and deeper weathering.

Nearly the entire country of Egypt—except for the Mediterranean coast and parts of the Eastern Desert and Sinai—lies in the zone of extreme aridity, receiving less than 25 mm of rain a year. In some areas, no rain falls for decades; then when rain does occur, it is usually in the form of brief, torrential storms. Such storms produce a large amount of precipitation over a very small area, and the runoff can be very erosive as it flows across the weathered surfaces. As we will learn in later chapters, the Tertiary Period was a time of much higher precipitation in Egypt than today. The Pleistocene witnessed the beginning of extreme dryness, but even then there were periods of increased rainfall called **pluvials**. Much of the present-day Egyptian landscape is the result of those wetter periods.

Wind can be a particularly effective agent of sediment removal in desert areas, where surfaces are dry and unprotected by vegetation. Wind-blown sand grains can also act as an abrasive to erode rock surfaces. Wind-blown grains seldom travel more than a meter or so above the ground, so erosion is confined to this zone. Water-born debris can also scour the surfaces over which the water is running. But if a weathered layer is not removed from a rock, it can actually act a barrier to prevent further breakdown.

The amount of precipitation, as well as the topography defining the drainage basin of rivers and streams, will affect the load of sediment that a river can carry. The greater the volume of water and the greater the velocity of the water, the more sediment and the larger the fragments that can be transported. The velocity of the river flow depends on both its volume and the **gradient** or slope of the river over the length of its course. Rivers in flood can move more sediment than normal flows. When the velocity of the water decreases because of flooding outside the channel, dispersal into a basin, or outflow into an ocean, the sediment load will be dropped.

The gradient of a river depends on the difference in elevation between its **catchment area** and sea level, although parts of the river may have gradients greater than or less than the average for the river as a whole. The gradient can be altered by tectonic uplift in the catchment area or by prolonged erosion that lowers the catchment area. Changes in worldwide sea level can also affect the gradient. Sea levels have changed many times, for example as a consequence of glacier formation and melting such as occurred several times during the Pleistocene Ice Ages, when vast amounts of water were locked up in the glaciers. Melting of such glaciers can cause a rather rapid increase in sea level. Other factors changing sea levels include greater volumes in mid-ocean ridges and continents colliding, reducing the amount of continental shelf.

A river with a low gradient or little volume will tend to deposit its excess sediment load on its channel bottom; such a river is said to be **aggrading** its channel. If the river is vigorous and picks up sediment in the course of its journey, it is eroding or **degrading** its channel. Study of sediments deposited by a river can provide information about past climatic conditions and drainage patterns based on the quantity of sediments, their size, and their composition.

To conclude this summary about the formation and disintegration of rocks, it is important to recognize that the surface of continents and their margins are undergoing continual change. New igneous and metamorphic rocks will form where continents are colliding, while sedimentary rocks will form along the coasts of continents. Rocks exposed or near the surface will undergo weathering processes. Weathering will be greatest in areas of low relief with high rainfall and temperature extremes, while high relief and high precipitation enhance erosion. Sediments, the products of weathering, will be transported by wind and water to lower levels, eventually being deposited in low-lying, water-filled basins inland or in the surrounding seas. Erosion will reduce mountains and plateaus to low plains; meanwhile tectonic forces may uplift other areas.

Dating Rocks

Throughout this book we will report that certain rocks were formed in a particular time period. How can a geologist date a rock? There are several methods—some of which work better with one kind of rock than another. The subject is extremely complex, and new methods are being devised all the time. The brief discussion that follows is only an introduction to this subject.

From the earliest days, geologists have used **stratigraphy** to establish a chronological ordering of rocks. This principle relies on the fact that younger rocks generally overlie older rocks. Exceptions to this have already been mentioned, but they can be recognized. This provides only a relative date for a rock: younger than the rock below it, older than the rock above it.

Geochronology provides methods of dating some rocks directly and (in theory) absolutely by studying the radioactive decay of certain elements within them. Each original (or 'parent') atom of a radioactive element decays to produce a characteristic 'daughter' atom of the same or different element. The ratio of parent to daughter atoms decreases with time and provides a measure of the time since the parent atoms were trapped in their rock. Errors may occur in this method if all the daughter atoms are not retained in the rock, as for example may easily occur with argon, the product of potassium-40 decay, which is a gas. For this reason, geologists may try to date a rock using several different radioactive materials. If different radioactive substances produce similar results, the conclusions are strengthened.

Among the three classes of rocks, only igneous and metamorphosed igneous rocks can be dated by this method. Radioactive dating cannot be used to determine when sedimentary rocks formed, since they contain fragments of much older rocks. Sedimentary rocks can often be dated, however, by examining the fossils in them. Because plant and animal species have undergone evolutionary changes, the kind of organic remains trapped in sediments varied with time. After many years of study, paleontologists have identified a set of **index fossils** that lived at known time periods and under known climatic conditions. Discovery of one of the index fossils in a rock sample immediately provides much information. By combining these methods geologists have managed to read and correlate the rock records throughout the world.

The Origin of Rock Types Found in Egypt

The premise of this book is that rocks and rivers have influenced the way in which successive Egyptian societies developed. We will see that there are many kinds of rock in Egypt, and some of them were very important to human activities. These include stone used for buildings, for tools, for art and jewelry, and as sources of minerals. The differing properties of rocks made them suitable for different purposes. For example, sedimentary rocks were generally preferred over igneous ones by ancient builders because such rocks split along their bedding planes and could be quarried more easily with the hand tools available. In some applications, however, the great strength of igneous rocks made them worth the extra effort that shaping entailed.

The specific location of useful rock within the countryside was also an important consideration. Rock needed in only small quantities could be sought out at distant locations, while that needed in volume for building large monuments could not be transported very far. In such cases, the monuments had to be sited close to the sources of stone. The placement of towns was primarily dictated by their relationship to the Nile and the desert, but proximity to certain kinds of rock or routes leading to mines or quarries could also influence their location. Furthermore, rocks were the ultimate source of the soil that supported agriculture. The presence of fertile soil plus the availability of water would determine the areas that could be cultivated.

Formation of the 'Rock Sandwich'

Geologists have developed a picture of the immensely thick rock foundation of Egypt by examining rock exposed at the surface or in the walls of the escarpments along the Nile Valley as well as by studying material brought up from deep bore holes drilled to locate water or oil. That foundation can be described as a sort of 'rock sandwich' with three principal components. From deepest and oldest to youngest these are: an ancient layer of igneous and metamorphic rock known as the **basement complex**, an intermediate layer of sandstone, and a top layer of limestone. Minor layers exist in addition to these, but they will be introduced gradually. Once the layers of rock were formed, erosional processes began to denude the surfaces, especially when they were exposed above sea level. Therefore, the rock that we observe on the surface of the ground (the so-called **bedrock** of an area) is a function both of the most recent kind of rock that formed there and of which layers of recent rock were removed so that older rock was revealed.

The basement complex is continuous beneath the entire country. It is only revealed at the surface in a few locations where uplift and erosion have combined to expose it; usually it is hidden beneath younger rocks. The basement complex is exposed in the far southwest at Gebel Uweinat, in scattered outcroppings in the southern part of the Western Desert, at Aswan in the first Nile cataract, in the high mountains of the Eastern Desert, and in the southern part of the Sinai Peninsula. The metamorphic rocks of the basement complex are tremendously old. Some of them have been dated by radioactive techniques to more than 2,000 million (2 billion) years. These occur in the southwest, where the ancient West Nile craton is located. Those in the Eastern Desert and Sinai are younger, with ages between 1,000 million and 500 million years. Many of these were formed during the period (estimated at 950 to 550 million years ago) when other plates were colliding with the African Plate. The metamorphic rocks are the oldest components of the basement complex, representing both sedimentary and igneous rocks that were transformed by the collisions. The igneous rocks are relatively younger than the metamorphic ones, having been intruded into the older rocks around them (See Figure 1.3).

Such ancient plate collisions may have produced a great mountain range. Tests of the underground surface of the basement complex show that it has only moderate relief, however, indicating that it probably underwent a long period of uplift and erosion before other rock layers of the 'sandwich' were deposited on top of it. We might also expect to find sedimentary rock formed from these ancient erosion products dating to somewhere in the Paleozoic Era. But such rocks occur in only a few spots in Egypt. Either these sediments were carried beyond the bounds of the current country or, as is more likely, the sedimentary rocks produced by early

17

erosion of the basement complex have long-since been uplifted and eroded completely away.

The sandstone layer in the middle of the 'rock sandwich' is usually referred to as **Nubian sandstone** because it is found in southern Egypt as well as in the northern Sudan—an area that was formerly known as **Nubia**. The Nubian sandstones have been dated to the late Cretaceous Period, to an Epoch around 80 to 90 million years ago. Ball (1912) described them as "consisting of medium-sized silica grains set in a more or less ferruginous cement." It is the ferruginous (iron bearing) cement that results in the yellow to red to brown colors. Manganese dioxide is sometimes present, giving the sandstone a purple to black hue.

A more recent description of the Nubian sandstone states: "The sandstones are highly porous and only loosely cemented with quartz, iron oxides . . . and chlorite clay. All three cements are almost always present . . . . Because it is incompletely cemented, the sandstone is friable and thus relatively easy to quarry and carve. Its strength and durability is imparted by the sparse but omnipresent quartz cement, and iron oxides give the sandstone its typical brown color" (Aston, Harrell, and Shaw, 2000:55).

The Nubian sandstones are exposed over a vast expanse in southern Egypt equal to about one-third of the entire country. Northward more recent layers still cover them. Above the Nubian sandstone, and also of Cretaceous age, are a layer of shale and one of fine white limestone (often called **chalk**). Over these layers was deposited another layer of shale during the Paleocene. These layers will be referred to simply as Cretaceous limestone/shale unless one of the sub-layers requires our special attention.

Finally, to top off the countrywide sandwich, an immensely thick layer of limestone was deposited during the Eocene. This limestone forms the exposed bedrock over the middle of the country as far north as Cairo. Apparently the rock foundation of Egypt was subjected to considerable compression in a north to south direction at the end of the Cretaceous Period. This resulted in considerable folding of the foundation on an east to west axis. During the following Eocene Epoch when the Tethys Sea again transgressed over the country, limestones were deposited in an ocean with a very uneven bottom. Thus the Eocene rock layers vary in thickness and composition at different locations. Because the Eocene deposition lasted for such a long time, during which conditions varied, a number of sub-layers can be distinguished. These details will be discussed as they become relevant.

The sandstones and limestones of the middle and upper layers of our 'rock sandwich' are both sedimentary rock types that can only be produced by sediments being deposited into an ocean basin. Egypt has been on the coastal margin of the African continent throughout the last several hundred million years. Evidently the Tethys Sea lying to the north of Egypt often extended farther south than the modern Mediterranean coastline and covered portions of the country. Whether this shift in coastline was the result of uplift and sinking of the Egyptian landmass or the result of worldwide changes in sea level, or more likely both, is immaterial. The sedimentary layers tend to be thicker toward the northern part of the country since that portion was always covered by the transgressing Tethys. All of these layers now slant gently downward toward the north as a result of later tectonic uplift of the southern part of the country.

These rock layers record a situation in which Egypt was repeatedly flooded by the sea and then exposed as dry land. Evidence that this extension of the sea occurred repeatedly, rather than continuously, comes from the fact that the sandstone and limestone layers reflect many different conditions of deposition with varying water levels over a given location sometimes deep, sometimes shallow and with the sediments reflecting different sources of parent rock. Heavy sediments such as gravel and sand are generally deposited in water close to the coast, so the coastline of the Tethys Sea must have varied back and forth from north to south very gradually to produce Nubian sandstone over such a vast tract. Although the sandstone contains few fossils, some floral remains have been found that suggest a warm and humid to semi-humid climate. Egypt at this time was probably just north of the Equator on a northward drifting African Plate. The shales were formed when the receding shallow seas were replaced by a series of lagoons and estuaries of brackish water. In these regions with slow-flowing water, silts and muds accumulated, trapping plant remains that washed in from nearby ridges. Limestones were formed during periods in which the land was covered by deep water, from which calcium carbonate precipitated, cementing together fragments of seashells. By examining the successive layers, geologists can recreate the ancient environments in which the rocks formed (Fig. 2.1).

Most of the country of Egypt received the rock layers just described, but only the extreme northern portion was submerged during the Oligocene and Miocene Epochs. During the Oligocene, the Tethys Sea coast lay between the

Agent	Deposition in Ocean			
Environment of Deposition	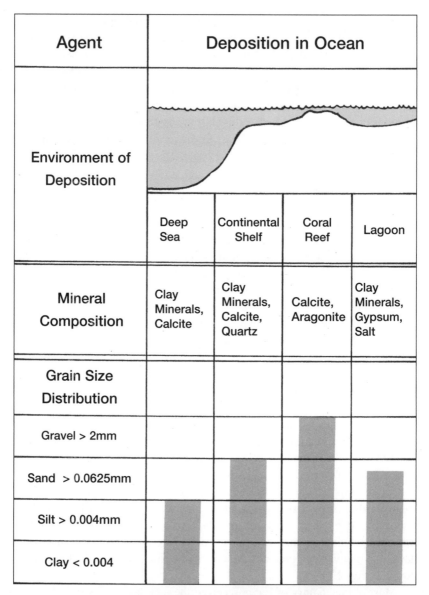			
	Deep Sea	Continental Shelf	Coral Reef	Lagoon
Mineral Composition	Clay Minerals, Calcite	Clay Minerals, Calcite, Quartz	Calcite, Aragonite	Clay Minerals, Gypsum, Salt
Grain Size Distribution				
Gravel > 2mm				
Sand > 0.0625mm				
Silt > 0.004mm				
Clay < 0.004				

Fig. 2.1: A summary of different rock types and their compositions when formed in different sedimentary environments.

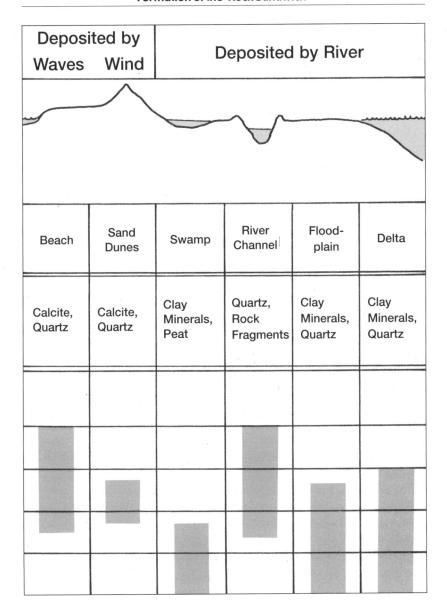

Deposited by Waves Wind		Deposited by River			
Beach	Sand Dunes	Swamp	River Channel	Flood-plain	Delta
Calcite, Quartz	Calcite, Quartz	Clay Minerals, Peat	Quartz, Rock Fragments	Clay Minerals, Quartz	Clay Minerals, Quartz

latitudes of the Fayum depression and Cairo. Many small rivers flowed toward the coast and carried gravel, primarily in the form of flint nodules eroded out of limestone, and dropped it in their channels. These conglomerate deposits also contain fossil remains of land animals and large tree trunks. Since this was a time when mammals were undergoing a rapid evolutionary radiation, these fossils are of great interest to paleontologists. During the early to mid-Miocene, another layer of limestone was deposited south of the modern Mediterranean coastline and along the shores of the Gulf of Suez.

Once the seas retreated (or more likely the land was lifted) permanently at the end of the Miocene, the forces of weathering and erosion began to act on the exposed rocks to 'peel back' the upper layers and expose deeper ones. Because the agents of denudation acted differently in different parts of the

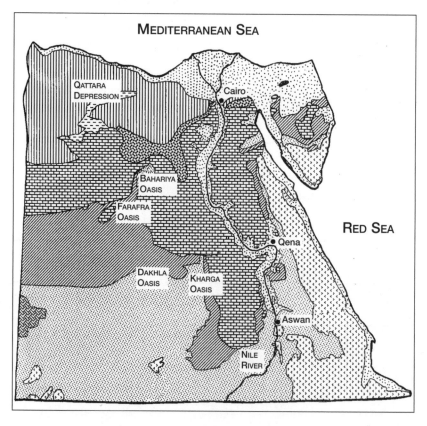

country, the rock layers were degraded to different depths. Thus rocks of various ages are exposed at the surface forming the local bedrock of their region (Fig. 2.2). These processes and their consequences will be described in greater detail in later chapters.

Two other kinds of rocks occur in certain regions of the country atop those already discussed. According to the principle of superposition, they are among the youngest rocks in the country. Extensive volcanism has occurred in some parts of East Africa cut by the East African Rift Valley. In

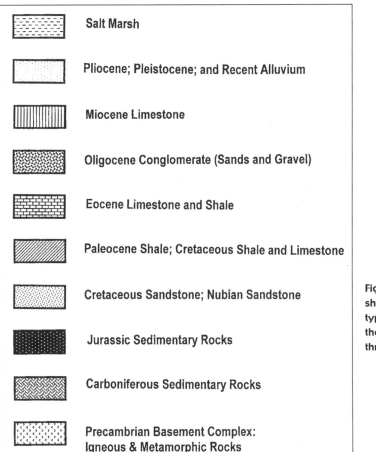

Salt Marsh

Pliocene; Pleistocene; and Recent Alluvium

Miocene Limestone

Oligocene Conglomerate (Sands and Gravel)

Eocene Limestone and Shale

Paleocene Shale; Cretaceous Shale and Limestone

Cretaceous Sandstone; Nubian Sandstone

Jurassic Sedimentary Rocks

Carboniferous Sedimentary Rocks

Precambrian Basement Complex:
Igneous & Metamorphic Rocks

Fig. 2.2: Map showing rock types found at the surface throughout Egypt.

Egypt, **basalt** was extruded from fissures to form sheets on the surface during the Oligocene. These sheets are found in the southwest at Gebel Uweinat and Gilf Kebir. They also occur as thin belts stretching from north of the Fayum to northeast of Cairo. The timing of these basalt eruptions suggests they may be related to other tectonic events of this period, namely the formation of the Red Sea.

The most recent rock layer might be called 'rock still in process.' This is the enormous quantity of sediments, or rock fragments, that have been transported from their parent rock by the River Nile or an ephemeral **wadi** stream, but which have not yet reached a final oceanic resting place. This material is called **alluvium**. It consists of gravel, pebbles, sand, silt, and clay. Some of this material has been distributed across the floodplain of the Nile Valley by the annual inundation, while additional piles of it still lie within the beds of wadis where it was dropped when the stream dried up or sank into the ground. We could include the sand sheets and dunes of the Western Desert and Sinai as another example of these recent but still moving deposits.

More information about these rock layers and why they appear at the surface in the locations where they do will be provided in upcoming chapters. The presence or absence of various kinds of rocks in the vicinity of certain settlements was to have important consequences for the Egyptians. We will also see that after the Nile Valley was cut deep into the bedrock throughout the length of the country, rocks at different depths became readily accessible.

Evolution of the Egyptian Landscape

By the end of the Eocene Epoch around 37 million years ago, several thick layers of rock covered the land of Egypt: from the bottom upward these were the basement complex, the Nubian sandstone, the Cretaceous limestones and shales, and the Eocene limestones. The land south of the latitude of Cairo was above sea level; but it was fairly flat and monotonous, with little or none of the current relief of high mountains, desert depressions, and winding Nile Valley. Soon, however, tectonic forces and geological processes would begin to create a varied terrain.

Formation of the Red Sea and the Red Sea Mountains

Geologists believe that the Red Sea was formed by the **rifting** of the Arabian Peninsula from the rest of the African Plate. This process began in the Oligocene and is continuing today. Several lines of evidence support this hypothesis. The Red Sea Mountains, the southern Sinai Peninsula, and Saudi Arabia, areas now separated by gulfs and the Red Sea, contain similar kinds of igneous and metamorphic rocks of about the same age; north of these rocks are limestone plateaus of similar ages with Eocene limestones as their top layer. This would most likely occur if the three regions comprised a single landmass during the formation of these rocks. The Red Sea is narrow, and its long sides are nearly parallel. Shifting the Arabian Peninsula to the west neatly merges the opposing shorelines and matches many rock formations. Along the axis of the Red Sea is a deep trench containing recently solidified magma; this trench is similar to those observed at mid-ocean spreading zones in other parts of the world.

The Red Sea is a good example of the formation of an ocean by the rifting of a continent: a process predicted by the theory of **plate tectonics**. As a

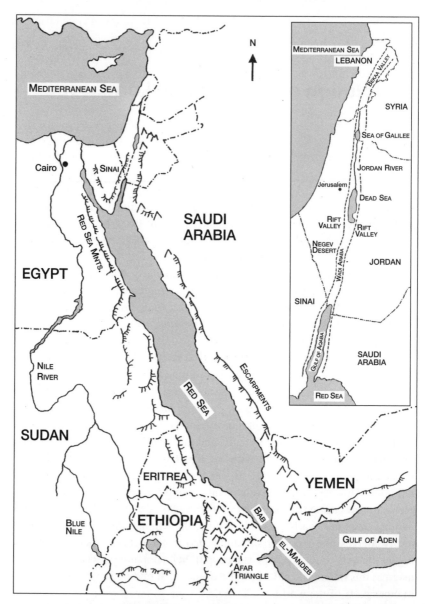

Fig. 3.1: The Red Sea, Gulf of Suez, and Gulf of Aqaba were formed by rifting of the African Plate. The rift continues northward into the Dead Sea Valley (inset).

result, geologists are studying the Red Sea with great interest, but many questions about its origin remain unresolved. The rift creating the Red Sea continues into the Gulf of Suez to the west of the Sinai Peninsula (Fig. 3.1). To the east of the Peninsula the rift continues into the Gulf of Aqaba, the Wadi Araba in the Negev Desert, the Dead Sea Rift system, and the Bekaa Valley in Lebanon. Portions of this rift valley already lie below sea level; at some time in the future they may become an extension of the Gulf of Aqaba.

To the south, the Red Sea rift is continuous with the large rift valley known as the East Africa Rift System. This system extends through Ethiopia, Kenya, and Tanzania for more than 5,600 km. A third branch of the rift system formed the Gulf of Aden; this gulf turns eastward at the south end of the Red Sea and joins it to the Indian Ocean. At the junction of the three branches of the rift system lies a most unusual region—the Afar Triangle. This area is largely below sea level, but coastal mountains prevent it from being flooded by the ocean. It has many volcanoes and briny lakes into which seawater seeps; the water evaporates from the lakes in the hot desert sun leaving deposits of salt.

The first stage of the rifting of the African Plate may have involved the formation of a dome, with the rock layers pushed up by hot magma rising from the mantle. Alternatively the crust may have been thinned and stretched by convection currents without the formation of a dome. Geologists are still debating the question of whether doming took place. Evidence seems to suggest that doming may have occurred at the southern end of the Red Sea but not at the northern end. The continental crust was too brittle to stretch very much; instead, cracks occurred and blocks of crust dropped, forming a valley. Then water from the Tethys Sea flowed in from the north and formed a bay. Finally the magma rising in the floor of this bay began to form new oceanic crust and this widened the Red Sea.

To the west of the Red Sea, the rock layers were uplifted to form the Red Sea Mountains. Similar mountains are found to the east of the Red Sea in Saudi Arabia, a fact that shows that rifting, sea floor spreading, and mountain building were somehow part of the same process. The mountains flanking the Red Sea rise as high as 3,000 m today, but they may have been much higher originally. In the wetter climates of earlier times, these mountains acted as catchments for atmospheric moisture, leading to increased rainfall. The runoff flowed down both the east and the west sides of the mountains and, encouraged by the steep gradients, began to erode the rock layers. The limestones and sandstones were easily eroded, leaving the more resistant igneous and metamorphic rocks of the ancient basement layer exposed in much of the range.

In addition to these mountains of ancient igneous and metamorphic rocks, new volcanic mountains have been formed in Ethiopia and Saudi Arabia by the upwelling magma always associated with rifting. Many volcanoes are found along the margins of the rift valley where it penetrates southwestward from the Red Sea into Ethiopia. Huge plateaus of lava have been generated over the years as the volcanoes erupted. The headwaters of the Blue Nile lie in these mountainous regions. The erosion of the lava produced the fertile sediments that were carried into Egypt by the annual inundation of the Nile and deposited over the fields. In Egypt, the erupting lava was runny and flowed from fissures over many square kilometers of the country rather then erupting and forming actual volcanoes.

After its separation from the African Plate, the Arabian Plate began to rotate counterclockwise and then to move northeastward until it collided with the Eurasian Plate, raising mountains in the region of Turkey and Iraq. This closed off the east end of the Tethys Sea, which until then had separated the African and Eurasian Plates, and created a narrow, nearly landlocked Mediterranean Sea.

Formation of the Nile Valley

Before the events of the early Oligocene described above, the landscape of northeastern Egypt was essentially level with sedimentary rock layers slanting slightly downward to the north. In the absence of a strong gradient, drainage patterns in this area were not well defined, and many shallow channels, rather than a single River Nile, carried sediments northward toward the coastline of the Tethys (or the Mediterranean Sea), which was probably in the vicinity of Cairo. During the uplift of the Red Sea Mountains, the gradient of these streams increased dramatically, and they became more erosive. Geologists call this process **rejuvenation**; we will encounter frequent examples of it. The streams carried a great quantity of gravel from the rising Red Sea Mountains and deposited it north of Cairo and northwest of the Fayum.

The uplift of the Red Sea Mountains in a range running north-northwest to south-southeast seems to have created a parallel sag or down-fold in the crust to the west. It seems reasonable to suppose that this sag plus the slight down trending of the rock layers toward the north directed the streams running westward off the Red Sea Mountains to divert toward the north along

28

what would be the future course of the River Nile. But it was to be another major geological disturbance that produced the Nile Valley we know today; that event was the complete desiccation of the Mediterranean Sea during the late Miocene Epoch.

When the idea that the Mediterranean Sea had once dried up was first proposed in 1970, it seemed an outrageous notion to many people. But as often happens in science, a bold new hypothesis suddenly united and explained a series of previously unrelated and puzzling observations. This hypothesis explained the presence of a thick layer of **evaporites** (rocks formed by precipitation from evaporating water) in cores drilled into the Mediterranean Sea floor, the detection of a deep and vast canyon underlying the Delta north of Cairo, and a buried valley under the Nile at Aswan. Deep but hidden valleys had also been found below rivers flowing toward the Mediterranean in Europe and North Africa.

It was not hard to find the geological cause of this catastrophic event. As the African Plate moved north against the Eurasian Plate, the floor of the Tethys Sea had been consumed by subduction, and the dimensions of the sea had been reduced to the area of the Mediterranean. This sea was eventually connected to the world's oceans only through the Straits of Gibraltar at its western end. While a number of rivers drain into the Mediterranean, so much water evaporates from the sea that its level is only maintained by inflow from the Atlantic Ocean. (Evaporation removes ten times as much water as enters from rivers.) In the late Miocene, either tectonic movements pinched off the straits or world sea levels fell below the shallow sill of the straits, or both, and the water of the Mediterranean evaporated (perhaps in as little as one thousand years according to Hsu, 1972) except for a few pools in the deepest basins. The Mediterranean Sea may have been dry for about a million years.

As the Mediterranean waters evaporated, the sea level fell, eventually exposing the abyssal plains more than 3,000 m below. The drop in sea level lowered the river's **base level** and thereby increased the gradient of the River Nile enormously. This increase in gradient combined with the large volume of runoff carrying huge amounts of sediment gave the river great erosive power. Within a few hundred thousand years, the enormous river—called the **Eonile**—carved a vast canyon into the layers of soft sedimentary rocks. This canyon stretched from the shore of the declining Mediterranean southward as far as Aswan and perhaps as far as Wadi Halfa. As the river cut deep into the rock layers, other erosion processes widened the valley. Some comparisons of this Nile Canyon with the Grand Canyon in Arizona may be helpful (Table 3.1).

Table 3.1: Comparison of late-Miocene Nile Canyon and Arizona Grand Canyon

	Nile Canyon (near Cairo)	Grand Canyon
Width	10 to 20 km	10 to 20 km
Length	1,300 km	320 km
Depth	2,500 m	2,080 m
Gradient of river	1:400	1:625

Based on Said, 1981

At the onset of the Pliocene Epoch around 5 million years ago, the Atlantic Ocean breached the Straits of Gibraltar and began to refill the Mediterranean basin—a process that may have required less than a thousand years. The seawater also invaded the Nile Canyon at least as far south as Aswan and extended into side valleys of the canyon, formed by tributary streams, to form inlets. While the canyon was flooded with seawater, marine sediments began to accumulate and fill the canyon to about one-third or one-half its depth.

While the Nile Canyon was essentially an arm of the Mediterranean during the early Pliocene, during the late Pliocene (beginning about 3.3 million years ago) the effect of a large river flowing from the south called the **Paleonile** became important. This changed the nature of the seawater in the canyon to brackish and then to fresh. It also changed the nature of the sediments that were deposited in the canyon: instead of seashells and other marine debris the rivers carried weathered rock fragments and terrestrial plant or animal remains from the Red Sea Mountains and the Western Desert. These riverine sediments finally filled the canyon to the rim, a rim cut in many places by the tributaries that flowed into the former Eonile. Later, erosion removed some of this sedimentary fill. The cliffs we see today on either side of the Nile Valley are the sides of the ancient Miocene Nile Canyon.

A long cycle of repeated periods of erosion and redeposition of sediments in the Nile Valley can be traced in the terraces that remain along the sides of the valley—above the modern floodplain—as well as in sediments now buried beneath recent alluvium. These record a series of aggradation and degradation events in which the Nile was responding to changes in its gradient and in the volume of water and sediments it carried.

Changes in the climate and rainfall within Egypt and beyond its southern border affected the volume of water in the river and its season of flow. Since rainfall also affects the amount of sediment produced in a river's drainage basin (as described in Chapter 1), the volume of sediment has varied. The Nile sediments reveal that at times the river was much larger (in terms of volume of water) than in modern times, while at other times it apparently stopped flowing altogether. Egyptian geologist Rushdi Said has analyzed the sediment data and defined a series of Niles as summarized in Table 3.2.

Table 3.2: Summary of series of late-Miocene and post-Miocene Niles (based on Said, 1993)

Dates of flow (thousands of years before present)	Geological epoch	Nile phase	Events
6,000–5,400	Miocene	Eonile	Mediterranean desiccation. Rainfall enhanced by Mediterranean evaporation. Nile sediments from Red Sea Mountains, whose heights led to increased erosion.
5,400–3,300	Pliocene	Mediterranean gulf	Nile Valley filled by seawater. Marine sediments half fill Nile Canyon.
3,300–1,800	Pliocene	Paleonile	Runoff and sediment sources local: Eastern and Western Deserts.
1,800–800	Pleistocene	Arid Phase	Nile stops flowing.
800–400	Pleistocene	Prenile	Vigorous river with water from Ethiopian sources and coarse sediments (sand and gravel).
400–12.5	Pleistocene	Neonile	Intermittent flows.
12.5–present	Holocene	Modern Nile	Ethiopian and Central African sources; flow perennial with summer flooding.

Sedimentary studies also reveal that the Nile has drawn its water from various sources. The river has apparently always carried runoff from the Red Sea Mountains and some parts of the Western Desert when enough rain fell on these areas. Today the river obtains most of its water from sources south of Egypt: the Nubian Desert, Sudan, the Ethiopian highlands drained via the Atbara River and Blue Nile, and the lake region of Central Africa draining via the White Nile (see next chapter).

The Ethiopian sources began to feed north into the **Prenile** only around 800,000 years ago, during the mid-Pleistocene. In sediments of that date one begins to find examples of minerals, plants, and animal remains that are characteristic of regions beyond the borders of Egypt. These sediments were composed of coarse gravels and sands, and they formed extremely thick deposits throughout the Nile Valley and Delta. These gravel and sand deposits are quarried for building material where they crop out along the edges of the valley. Elsewhere across the floodplain more recent deposits of silt and clay cover them. This highly-porous Prenile sand is trapped between two impermeable layers of clay and acts as an excellent reservoir of ground water. Wells dug to access this water have to be at least 40 m deep. The water is abundant but contains a lot of minerals. Cairo has long used this as a source of water—a resource that became especially valuable during cholera epidemics, when the water in the Nile could have been a source of infection.

The gigantic Prenile River flowed for several hundred thousand years. Then it dwindled to a smaller watercourse that varied in its volume and sometimes disappeared altogether. Among the sediments of the late Pleistocene are enormous quantities of silt that the river deposited in Nubia and southern Egypt. It has been proposed that these silts came from a 'Lake Sudd' that lay in southern Sudan where today a vast swamp is located. For thousands of years, the Sudan lake received water from the south but had no northern outlet to the Nile. Then a stream broke through the northern rim of the lake and connected it to the north-flowing Nile. The silt that had accumulated in the lake was washed downstream into Nubia and Egypt. The upper, most recent layers of the silt contain flint tools fashioned by prehistoric humans, and these have enabled geologists to date the silts to known **Paleolithic** periods. About 12,500 years ago, the modern Nile with its characteristic summer floods came into existence. Its behavior since then, especially in historic times, and its impact on the Egyptian civilization will be discussed in the next chapter.

In recounting the six-million-year history of the Nile, Rushdi Said notes that changes in the earth's crust and climatic fluctuations affecting the entire globe as well as Africa have had major impacts on the Nile and its Valley. There is no reason to think that global changes will not continue. Said argues that it is possible that the connection to the southern sources of the Nile's water could be broken again in the future, as it has been in the past, if tectonic forces raise the northern edge of the Sudan basin. Climatic changes of either a temporary or permanent nature could also have a devastating effect by reducing the flow of Nile water on which 70 million inhabitants of Egypt, not to mention 32 million in the Sudan and millions more people in other riparian countries, depend.

Nile Tributaries

Today the River Nile receives no major amounts of water from tributaries on its 2,700 km course through lower Sudan and Egypt to the Mediterranean. This was not the case in the much wetter time of the late Miocene, the Pliocene, and certain periods of the Pleistocene. Instead, many streams entered the Nubian and Egyptian Nile from the east and west and made a major contribution to its flow. Today these former tributaries survive as only dry streambeds called **wadis**. A wadi is a dry desert valley whose cliff-like walls rise abruptly from the flat floor. While few wadis have water flowing in them permanently, after a local rainstorm they can be raging torrents. The water flows in a variable channel and in spite of its speed cannot move the huge amount of debris far before seeping into the valley floor or evaporating. Therefore, large quantities of weathered rock collect on the floor of a wadi. Where wadis open into a floodplain or into an inland basin they drop their sediments and form **alluvial fans**.

The presence of a wadi mouth and the length of the wadi as it leads into the adjacent desert were important geographical considerations for the ancient Egyptians. Towns were often set near the mouth of a wadi that formed a convenient route through the desert, and alluvial fans offered an elevated site for building above the floodplain. Early farmers sometimes planted crops in the wadi alluvium after a storm.

The Nile from Historic Times
to the Present

The Nile is the longest river in the world. Its tributaries tap a variety of sources of water in a **drainage basin** covering nearly 3 million sq km and including parts of nine African countries besides Egypt: Burundi, the Democratic Republic of the Congo, Eritrea, Ethiopia, Kenya, Rwanda, Sudan, Tanzania, and Uganda (Fig. 4.1). In spite of this vast catchment area, the discharge of the Nile at its mouth is only 15 percent that of the Mississippi, with a comparable drainage area (Table 4.1). Even a river such as the Danube, which drains a much smaller region, has two and one-half times the discharge. The problem is that the River Nile flows for the final 40 percent of its length through deserts from which it obtains no water and in which there is significant evaporation and seepage. Furthermore, much of the water that falls in the equatorial upper portion of the drainage basin evaporates before it reaches the Egyptian Nile and the Mediterranean. By contrast, the River Amazon, with a length similar to the Nile and a drainage basin only about twice as large, has a discharge 66 times as great because it lies in the equatorial region in which there is heavy rainfall over its entire basin.

Table 4.1: The Nile compared to some other major rivers of the world (based on Said, 1993)

River	Length (km)	Drainage area (sq km)	Annual discharge (billion cubic meters)
Nile	6,825	2,960,000	84
Mississippi	3,766	3,270,000	562
Danube	2,900	816,000	206
Amazon	6,700	7,050,000	5,518

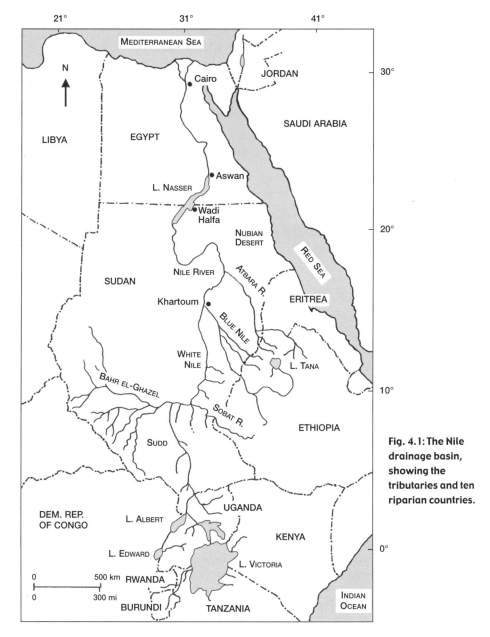

Fig. 4.1: The Nile drainage basin, showing the tributaries and ten riparian countries.

The discharge value in Table 4.1 does not tell the whole story of the Nile's flow historically, however, because the annual discharge was not spread evenly over the year. Most discharge came during the annual flood occurring from August to October. This is different from many rivers whose flood season is spring, as snow in the upper elevations of their drainage basins melts. The Nile had a late summer and early fall flood season, and this had a major bearing on the agricultural cycle of Egypt (Fig. 4.2).

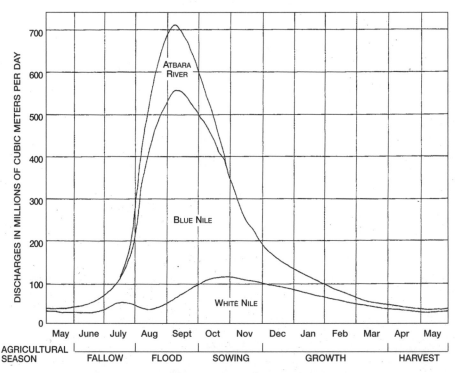

Fig. 4.2: The pattern of the annual discharge of the Nile averaged over a number of years before the construction of the Aswan High Dam. The contribution of each tributary toward the total discharge (top line on graph) is indicated. The flow of water from the White Nile was retarded in August and September by the huge influx from the Blue Nile. Activities of the single-crop agricultural cycle are shown below the graph.

The Nile Drainage Basin and Flooding

The White Nile is one of the three main tributaries of the Egyptian Nile. It has its origin in central or equatorial Africa, where the rainfall runoff is collected in a series of lakes. Lake Victoria is fed primarily by rainfall throughout the year on its huge catchment area while Lake Albert also receives runoff from the Ruwenzori Mountain Range, one of the wettest places on earth. Much of the water contributed by these lakes, however, evaporates in the spreading swamps of southern Sudan known as the Sudd. At the northern edge of the Sudd, the White Nile receives additional water from the Bahr el-Ghazal and the Sobat River. The White Nile has a fairly even flow of water throughout the year, increasing only slightly in the fall as a result of the Sobat's increased flow at that time.

The other two main tributaries of the Nile—the Blue Nile and the Atbara River—drain the highlands of Ethiopia. These rivers fluctuate enormously in volume as the monsoon summer rains swell them manyfold. It is this seasonal deluge that created the annual flooding of the Nile within Egypt. Variations in the height of the flood were caused primarily by variations in the rainfall over Ethiopia; variations in the White Nile's volume had very little effect in Egypt. At the present, the Egyptian Nile receives little or no water from the Nubian Desert of northern Sudan, although deep wadis there indicate that in prior wet periods much water flowed to the Nile from this region.

The Nile floodwaters carried vast amounts of suspended material, especially silt and clay, eroded from the volcanic rock in the mountains of Ethiopia. This material and various chemical constituents made the sediments extremely fertile. When the floodwaters exceeded the capacity of the river channel, the waters spread out on both sides across the **floodplain**. The Nile Valley is bounded on both sides throughout most of its length from Aswan to Cairo by rocky cliffs and terraces of gravel; these formed outer limits to the distance the floodwaters could move. In some places the floodplain is bounded by sandy desert with only a slight elevation. The edge of the floodplain can still be clearly identified, however, as the lush green changes to dry brown.

When fast-flowing floodwaters leave a river channel to spread across the floodplain, the depth of water decreases and the velocity of flow drops. The reduced velocity reduces the carrying capacity of the water, so the coarsest sediments settle immediately adjacent to the river. These sediments quickly accumulate and form ridges parallel to the river called **levees** that rise from one to three meters above the floodplain. The finer sediments in suspension

37

are carried farther and settle gradually at greater distances from the river. As a result, the cross-section of the floodplain has its highest points along the edges of the river channel (as well as along abandoned river channels); the plain slopes downward into **basins** at the edges of the plain.

Egyptian Agriculture: Flooding and Irrigation Practices

The annual flooding of the Nile was an event of supreme importance to people living in the Nile Valley. It controlled the calendar of activities of the agrarian society of ancient Egyptians. They recognized three seasons: flooding (or inundation), sowing and growth, and harvesting. The inundation distributed a fresh layer of silt and clay that maintained the fertility of the land. It was the moisture, however, that was essential on a yearly basis. Only land that had been covered by the floodwaters could be cultivated that year. Under this regime, only one crop per year was possible in the basins, and it consisted of such 'winter crops' as barley, wheat, and flax.

The practice of allowing the flood to fill the low areas of the floodplain is referred to as basin irrigation. 'Irrigation management' took the form of augmenting natural levees and constructing dikes perpendicular to the levees to divide the floodplain up into basins. The exact date when basin irrigation was instituted in Egypt is unknown. Tradition attributes it to King Menes, who is supposed to have united Upper and Lower Egypt around 3100 BCE (see Table 4.2). But there is evidence that irrigation on at least a local scale was already in use by the end of the Predynastic period. A mace head of Predynastic King Scorpion shows him with a hoe about to breach a dike. The conversion of the entire floodplain into managed basins was gradual, with additions to levees and dikes proceeding from Upper Egypt along the river northward. (Note: because the Nile flows from south to north, Upper Egypt is in the southern part of the country, while Lower Egypt is in the northern part.)

Floodwater was diverted through the levees via **feeder canals** and introduced into the basins, where it remained for 40 to 60 days in order to thoroughly saturate the soil and let the fine sediments settle. Water that did not soak into the field during that time was drained back to the river through drainage canals—although in years of low flooding, water from one basin might be channeled into another basin. For the majority of the fields, no lifting or pumping of water was done since no adequate technology existed for that purpose until the New Kingdom and even later.

Table 4.2: Chronology of Egyptian history		
Period	*Dynasties*	*Dates*
Predynastic	Neolithic, Dynasty 0	5500–3050 BCE
Early Dynastic	I–II	3050–2687
Old Kingdom	III–VI	2687–2191
First Intermediate Period	VII–X	2190–2040
Middle Kingdom	XI–XIV	2061–1665
Second Intermediate Period	XV–XVII	1664–1569
New Kingdom	XVIII–XX	1569–1081
Third Intermediate Period	XXI–XXIII	1081–711
Late Period	XXIV–XXXI	724–333
Ptolemaic Period		332–30
Roman/Byzantine Period		30 BCE–641 CE
Islamic Period		642–1517
Ottoman Period		1517–1914
British Protectorate		1914–1952
Egyptian Republic		1952–present
Up to Roman period, based on Redford, 2001.		

A normal flood would provide a depth of about 1.5 m of water in each basin. Extreme variation in the height of the flood could presage a year of misery and famine. Floods below normal level would not rise high enough to cover all the basins and would not stay on the fields long enough to thoroughly saturate them. Reduced harvests would be inevitable. Floods of much greater than normal heights were equally unwelcome. The greater volumes of water would flow over the levees destroying houses, trees, vineyards, and gardens. People and animals could be drowned. The rushing water would breach the earthen dikes and wash away **sluice gates** along the canals. Years of hard labor would be required to restore these structures. The greater volume of water would linger in the fields so long that planting would be delayed. Not until the nineteenth century CE would technology exist to allow Egyptians to control the floods to some degree, and not until the construction of the Aswan High Dam in the 1960s would Egypt be completely free of the threat of devastating floods.

The basins could not be used for summer crops, since during summer months they were either lying fallow or under water. The only areas available for planting of trees, vines, and summer vegetables were the raised levees on each side of the river. These levees were continually augmented by the Egyptians to make them higher and wider—making them less subject to breaching by floodwaters and providing more area for cultivation and habitation. Initially the only source of water for these plots was the river or wells dug into subsoil water.

During the Old and Middle Kingdoms, all water had to be moved in pots or buckets by men or perhaps donkeys. Since the levees were (typically) never covered by floodwaters they did not receive the nourishing silt. They had to be artificially augmented and fertilized. Pigeon droppings collected from the large dovecotes were used for this; other forms of animal manure were too valuable as fuel. Nitrogen-enriched soil from crumbling mud-brick buildings (called **sabakh**) was also sought after as an amendment.

Later, devices to lift water were invented or introduced that made the growth of two crops per year on these raised places more practical. The **shadoof** consists of a long pole pivoted on a tall post. It has a bucket at one end and a counterbalancing weight at the other end. It is worked by hand by lowering the bucket into a water source; then the counterweight lifts the bucket so that water can be emptied out at a higher level. The shadoof was employed at least as early as the New Kingdom, when it appears in tomb scenes, but its irrigating potential was limited. It has been estimated that two men using a shadoof could irrigate about 1.5 hectares (four acres) of land. The Archimedes screw was introduced into Egypt in the fifth century BCE, while introduction of the *saqya* or waterwheel occurred slightly later in Ptolemaic times. These devices were operated by animals. A pair of oxen working a *saqya* could irrigate about 5 hectares. A waterwheel could not only lift more water than a shadoof, but it could lift it two to four times higher or up to six meters.

Karl Butzer estimates that from the Predynastic period through the Middle Kingdom, about 8,000 sq km of land in the Nile Valley were under cultivation using the basin irrigation method. Double cropping on the levees using first the shadoof and then the *saqya* each increased the cultivated land area by 10 to 15 percent. By 150 BCE, he estimates the total arable land throughout Egypt to have been 27,300 sq km, of which 10,000 sq km were in the Valley, 1,300 in the Fayum, and 16,000 in the Delta. This is nearly equal to the area farmed in the 1880s. Very little had changed in the irrigation technology in two millennia.

Variations in the height of the floodwaters from year to year were of considerable concern to the Egyptian government. Several successive years of low floods could bring famine, since only a limited amount of surplus food could be grown in good years to be stored for lean ones. From the beginning of dynastic history, heights of floods were measured and recorded. Some ancient documents have been found such as the Palermo Stone carved in the Fifth Dynasty that give a year-by-year account of flood levels. More indirect evidence of agricultural conditions has come down to us in other kinds of writings offering eyewitness accounts of the social and economic conditions. Study of river terraces and sediments is still another source of information about the past behavior of the Nile. These records provide evidence of short-term fluctuations around a mean as well as long-term trends to higher and lower floods.

Several authors (see Bell, 1971; Butzer, 1976; and Said, 1993) have studied these records in detail. Although the pattern is far from simple, there appears to be a strong correlation between normal flood levels and national prosperity and stability. For example, after recording adequate or above-average floods for the First Dynasty and most of the Second Dynasty, the Palermo Stone indicates that flood levels were below average at the end of the Second Dynasty, a period of known civil discord. Flood levels increased generally and showed less variation in the Old Kingdom—a period of great prosperity characterized by pyramid building, great art works, and elaborate tombs of officials.

During the Sixth Dynasty and the following First Intermediate Period, Nile flood levels apparently fell to low levels and remained there for an extended period. During the First Intermediate Period, there was a complete breakdown in the control by the central government in Memphis and much social unrest. Power was exercised by local officials if they were capable. Many causes of the breakdown have been suggested including a decline of the king's power during the 97-year reign of Pepi II, last king of the Sixth Dynasty, and the growing independence of local officials, but any king or bureaucrat would have been powerless to avert the economic decline caused by persistent low floods. No official flood level records exist for this period, since the central government was in disarray, but conditions have been deduced from many personal letters and individual documents that have been discovered. These documents mention many situations that would follow from several decades of low flooding: crop failures and famine, civil disobedience, sandstorms, and even perhaps cannibalism.

This sustained series of low floods (postulated by Bell to have lasted from 2180–2130 BCE) was not simply a random string of low years but reflected a fundamental change in climate and the end of the latest Holocene **pluvial**, the so-called **Holocene Wet Phase**. This wet phase began around 10,000 years ago, concurrent with the beginning of the Epoch known as the Holocene. At that time a climatic zone that produces from 10 to 50 cm of precipitation a year across sub-Saharan Africa moved northward. Increased rain fell over a band extending across Africa, turning the region that is today the arid Sahara into a savannah land. The increased rainfall also fed several Nile tributaries (especially those of Nubia, northern Sudan, the Eastern Desert, and the Atbara), increasing the river's flow and the heights of its floods. For the next 6,500 years conditions fluctuated, with several wet phases of a thousand-year duration separated by drier phases of lasting around 500 years. Evidence of these phases can be found in the stratigraphy of the Fayum depression south of Cairo and of the Nabta Playa in the Western Desert. When the Holocene Wet Phase ended, the climate of southern Egypt became hyperarid, as it remains today. The savannahs gradually became deserts, and the River Nile was much reduced in volume. From then on its floods would vary about a much lower average level, and a sustained period of below-average floods would bring disaster by disrupting agriculture.

Adequate floods were again the norm during the Middle Kingdom, when a strong central government regained control of the entire country and ushered in a period of prosperity and progress. Above average floods occurred during the Twelfth Dynasty and apparently did great damage to the area around Memphis and the Delta. This prompted King Amenemhat III to approve a scheme to divert excess floodwaters into the Fayum depression. He had the natural channel leading from the Nile into the depression (the Hawara Channel) widened and deepened and fitted with sluice gates. Water amounting to up to 10 percent of the flood discharge was admitted to the depression during peak flood and returned to the Nile (less that which had evaporated) after the crest of the flood had passed. The level of the lake in the Fayum was controlled at a maximum of 21 m above sea level, and almost 85 sq km of land near the entrance to the depression lying above this level was prepared for cultivation. The period of higher than average floods (estimated at 1840–1770 BCE) almost certainly reflects a temporary change in climate, reverting to a situation similar to that of the Holocene Wet Phase. But this improvement was not to last.

Few records exist to provide a basis for correlating flood levels with the events of the Second Intermediate Period—another time of government breakdown and foreign conquest in the Delta. Flooding was apparently nor-

42

mal or slightly better for most of the New Kingdom, but declined during the Twentieth and Twenty-first Dynasties, which may have been a factor in the general recession seen in those dynasties. This summary of the events during the flowering of Egyptian civilization is sufficient to demonstrate the sensitive relationship between Nile flood levels and political stability.

Records about flood levels continued to be kept with more or less fidelity into the modern era, but many of these have been lost and the others are of uncertain value for making comparisons, since the basis for their measurements varied as the riverbed rose and as calendars changed. Certainly reports of famines continue to occur throughout Egyptian history, showing that fluctuations in flood levels probably continued to be a problem beyond the control of the national government. Even in times without strong central control, it behooved local villagers to maintain the dikes and levees, since the proper channeling of the water to their fields depended on this and was unrelated (except in the general sense of high or low flooding) to what was occurring upstream or downstream. For this reason, geographer Karl Butzer has challenged the hypothesis of some Egyptologists that a need for centralized control of Nile irrigation was an important factor leading to the evolution of the Egyptian monarchy. Instead he concludes that irrigation was always under local control, although taxes were set and collected on a national basis. Agricultural produce was also collected and redistributed primarily by local institutions, especially temples, which had royal patronage.

Damming the Nile

For 5,000 years, throughout dynastic history up to the nineteenth century, Egyptian agriculture continued to depend on natural flooding to provide the water for the single winter crop and on primitive water-lifting devices to water the raised lands on which two crops would be grown. In the early nineteenth century Muhammad Ali, the Egyptian viceroy, decided to plant huge fields of cotton, a summer cash crop that required water just when the river was normally at its lowest. His plans led to new techniques to manage the Nile's water. In addition to obtaining water during the summer months, it was also considered desirable to control the flood to make agriculture less subject to the variations in flood level, to increase the area of cultivated land by better use of the flood waters, rather than letting much of it run to the sea, and to increase the lands on which two or even three crops per year could be raised, by instituting a system of **perennial irrigation**.

Several unsuccessful methods were attempted to achieve these goals, until finally the first dam (called by the Egyptians a **barrage**) was built across the river at the apex of the Delta about 20 km north of Cairo, where the river divides into the Rosetta and Damietta branches. The objective of the barrage was to raise the water level behind it high enough to fill three feeder canals. These canals ran along the high levees of abandoned **distributaries**; thus their water was high enough to flow into the fields by gravity and then drain off into the two Nile branches. This dam did not store floodwaters but simply impounded the river's flow from February to July to a level 4 m higher than usual. During the height of the flood, sluice gates in the dam were opened to allow the silt-ladened floodwaters to flow through. This prevented the silt from collecting behind the dam.

The first effort to store floodwaters behind a dam and then release the water back into the river at times of low flow occurred in 1902, when the first dam was built at Aswan. This dam was twice increased in height, in 1912 and 1934. The dam was designed let the peak flow of the floodwaters carrying their heavy load of silt to pass through; then it impounded water from November to January. The first rush of floodwaters irrigated the basins for a winter crop in the traditional manner. The stored water was released to supply the crops from February to July. Later, barrages were constructed at several other points along the river between Aswan and Cairo (at Asyut, Zifta, Esna, and Nag Hammadi). These dams raised the river level by 3 or 4 m at each location, allowing other feeder canals to leave the river just upstream of them to supply water for perennial irrigation in fields along the river. A dam was also built in 1937 on the White Nile 40 km above Khartoum in Sudan and paid for by Egypt to increase the amount of water available for Egyptian summer crops.

The Aswan High Dam

The dams already described served the function of impounding part of the floodwaters of one annual flood and releasing them during the period of low water that immediately followed. There was no attempt to store all the floodwater or to carry water over from one year to another in order to average out the flows from different years. Both of these objectives were considered high priorities to improve Egyptian agriculture and to provide food, energy, and manufacturing for a rapidly growing population. Throughout the early decades of the twentieth century many ways of achieving these goals were considered by the Egyptian Ministry of Irrigation.

The earliest schemes were based on the premise that most of the countries within the Nile drainage basin were under British control and would be amenable to projects within their borders that would primarily benefit Egypt and Sudan, which at that time were the only countries using the Nile's water for irrigation. Thus early plans called for creating storage reservoirs at Lakes Victoria, Albert, and Tana and for digging a canal through the Sudd to improve the flow of water and reduce evaporative losses. After the Egyptian revolution of 1952, consideration focused on schemes that would permit direct Egyptian control. And as the countries of the upper Nile basin gained their own independence, it became clear that issues of national sovereignty would not allow them to accept an 'Egypt-centered' development plan. Finally, Egypt realized that only the construction of a massive 'high dam' at Aswan would ensure its water requirements without being subject to other countries' political instabilities or competing designs to use Nile water. The High Dam was completed in 1970, creating the huge reservoir of Lake Nasser behind it.

Consequences of the Aswan High Dam

Because the High Dam impounds all the floodwaters, flooding has ceased on the floodplains north of Aswan; instead the fields are irrigated on a regular basis at times and in amounts adjusted to the needs of various crops. The water level in the river is maintained at a level between the old low water and flood stages—a level that is high enough to permit water to be drawn off into canals for irrigation. This higher level also permits navigation on the river at all times of the year, and river traffic has increased as a consequence. The conversion to perennial irrigation means that two or sometimes three crops can be grown on most cultivated acres. Data about the actual increase in acreage cultivated have been hard to obtain, but there has been some increase, which has helped to compensate for lands removed from cultivation by urban sprawl and silt digging to make bricks. Not all areas scheduled for irrigation have turned out to be suitable farmland.

Almost immediately, the High Dam achieved its objective of balancing out the years of high and low flooding. During the 1970s and 1980s the rainfall over central Africa was less than average, and Egypt was able to meet its irrigation needs by drawing down the Lake Nasser reservoir. During this period, countries south of Egypt were devastated by droughts and famines. Equally important, Egypt is no longer under the threat of devastation from higher than normal floods, such as the one that occurred in 1975 that was the third largest flood on record. Such floods would now do incredible damage. Electrical power generated at the dam has supplied as much as 53 percent of the country's needs, but is averaging around 33 percent.

The higher water level in the river results in a higher **ground water level** throughout the length of the Nile Valley. The ground water level used to fall to 3 or 4 m below the surface during the early summer, but now the water level is around 2 m higher than before in Upper Egypt. In the past, the draining of the floodwaters from the basins also flushed out salts. Since this no longer occurs, the soil is becoming too salty. At the same time, the land no longer receives its annual layer of fertile silt and requires synthetic fertilizers to maintain productivity for the two or three crops harvested from each field. It should be noted, however, that the conversion of much of the arable land to perennial irrigation between 1900 and 1970 meant that most fields were already deprived of the annual silt layer, so the increased use of fertilizers cannot be blamed entirely on the High Dam. Geologist William Hume was already warning about saturated and salty soils in 1925. Pesticide use has also increased as the constant humidity and standing water allowed parasites and insects to flourish. The fear that *Schistosomiasis* infections would increase has not been realized, however, thanks to improvements in village drinking water supplies, health education, and medical care.

It now appears that none of the environmental disasters predicted before the dam was built has occurred. And although some negative consequences have been encountered, most people would probably feel now, as indeed most experts did before dam construction, that the benefits have outweighed the costs. The major objectives have been met: flood control, conversion from basin to perennial agriculture, and energy production. The major negatives of water saturation and salt buildups in fields should be solved by better drainage and more efficient application of irrigation water. Continued vigilance is necessary, however, to monitor these and other possibly unforeseen consequences.

Lake Nasser and Nubia

With this chapter, we will begin an imaginary trip down the River Nile. As we cruise along the river, we will see how the geological principles presented in the first few chapters apply to specific settings. Egypt south of Aswan, together with a part of northern Sudan, comprised the ancient land of Nubia. This desert land was sometimes dominated by Egypt, while at other times it pursued an independent course, and once, in the Twenty-fifth Dynasty, a Nubian family occupied the throne of the Egyptian pharaohs. Nubians traditionally lived in small villages along the river, where they pursued agriculture, fishing, and some herding. They were also involved in the extensive trade between Egypt and central Africa that passed along the river.

Formation of Lake Nasser

Aswan (93 m above sea level) was the logical place to build dams across the Nile, since boats could not ascend farther because of the cataracts and since the floodplain narrowed significantly along the Nubian stretch of the river. The first three dams at Aswan (actually one dam enlarged two times) were only intended to store the water of one flood season and release it at the next low water. They created reservoirs with increasing heights: the 1902 dam raised a lake to 106 m above sea level, the 1912 dam to 113 m, and the 1934 dam to 121 m.

The High Dam at Aswan was completed in 1970, and since then the water of the Nile has collected to form the 500 km long Lake Nasser (Fig. 5.1). The lake averages about 10 km in width, but bays extending up into former wadis make its outline very irregular. When the lake is filled to its maximum level of 175 m above sea level, the reservoir holds nearly 140 billion cubic meters (30 cubic miles) of water. The level rises and falls as floodwater is retained and then released over the course of the year.

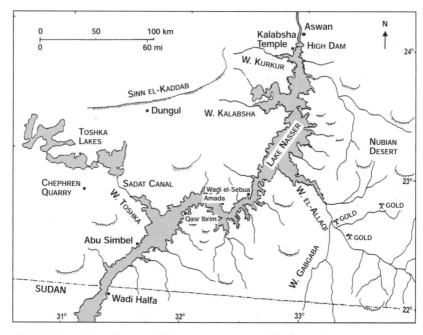

Fig. 5.1: Map of Lake Nasser, with key archaeological sites. The position of monuments from Kalabsha, Wadi el-Sebua, and Amada is shown after their relocation. W = Wadi.

From 1900 to 1960, the yearly flow of the Nile averaged about 84 billon cubic meters of water, but there was considerable year-to-year variation. Based on this historical average, a 1959 treaty allotted Egypt 55.5 billion cubic meters per year from Lake Nasser and Sudan 18.5 billion; the final 10 billion cubic meters was allotted to evaporative losses. It was recognized that if inflow exceeded this average for too long a period, the water level would exceed its safe limit. This led to the addition of an overflow canal (to be described below). In fact, there was an exceptionally high flood in 1975 of 125 billion cubic meters, but this was easily accommodated since the reservoir was still filling. By contrast, flows in three years were abnormally low: 1972 had 60 billion, 1983 had only 40.7 billion, and 1987 had 48 billion cubic meters. Water from the reservoir made up for the shortfall. Clearly, in these years the lake served its purposes of preventing catastrophic flooding or a shortage of water.

One disadvantage of the Aswan location for the High Dam is that the dry desert air evaporates a great deal of water from the surface of Lake Nasser. Yearly evaporation and seepage losses average about 10.8 percent of the volume of water in the reservoir. Seepage is expected to decline, however, as neighboring aquifers become saturated.

Since the dam traps all the floodwater, it also traps the suspended materials carried by the flood. This amounts to around 110 million tons per year. The coarse sediments collect at the south end of the lake in the vicinity of the Second Cataract, where the lake is already shallow. These sediments have accumulated to the point of emerging as islands above the surface. The designers of the dam anticipated that silts and clays would accumulate, however, and allocated 20 percent of the lake's volume to hold them. This provision should accommodate the sediments for 400 to 500 years.

Limited agriculture had been planned for the lakeshore, but the soil was found to be unsuitable and the fluctuating water levels a problem. A fishing industry has been established on the lake, however, that provides a source of income for some Nubians who settled in Aswan. The natural vegetation that has grown along the elongated banks has attracted nomadic herders from the Eastern Desert.

The scenery along Lake Nasser is less dramatic than that of the former Nubian Nile, but one has a better view of the landscape beyond the river. The surface of the desert on each side of the Nile is a plain with an elevation of around 20 m above the lake, or around 200 m above sea level. This plain is deeply incised with wadis entering from both the east and the west. It is easier to appreciate their ramifications from the air than from the deck of a cruise ship. Beyond the first plain are two others stepping back to the west and east at elevations above sea level of 230 to 260 m for the first stage and 300 to 360 m for the highest level. The surface of the two lower plains is made of Nubian sandstone, while the surface of the highest plain (consisting of younger rock) is composed of Cretaceous limestone and shale. To the far west of the lake an escarpment of Eocene limestone, the Sinn el-Kaddab, marks the southeast edge of the vast limestone plateau that separates the oases of the Western Desert from the Nile Valley.

These three plain surfaces at successively higher elevations are termed **pediplains** and were formed by earlier cycles of erosion. A pediplain is an erosional surface characteristic of arid environments. Its formation can be explained as follows. Suppose one begins with a plateau at some elevation above sea level. Rain falling on the plateau will flow into streams that will

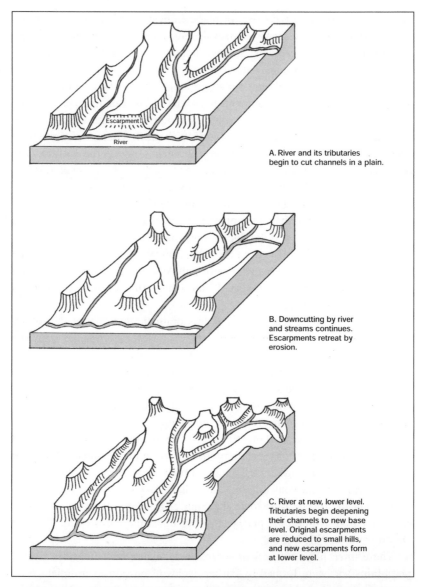

A. River and its tributaries begin to cut channels in a plain.

B. Downcutting by river and streams continues. Escarpments retreat by erosion.

C. River at new, lower level. Tributaries begin deepening their channels to new base level. Original escarpments are reduced to small hills, and new escarpments form at lower level.

Fig. 5.2: Diagram illustrating the sequence of events involved in the formation of a pediplain landscape. Such landscapes are characteristic of arid environments but must have begun their development when rainfall and runoff were abundant.

begin to cut channels into the surface of the plateau (Fig. 5.2). These channels will gradually widen until the plateau is dissected into a series of isolated **buttes** surrounded by a plain whose elevation is equal to that of the **base level** of the drainage—either sea level, or in this case since we are far inland, the level of the river into which the wadis drained. The isolated buttes will continue to erode by undercutting, rock falls, and wind deflation.

If the base level falls, another cycle will begin with downcutting of the plain formed in the first cycle. (The same effect can be achieved by a tectonic uplift of the plateau region.) After several cycles of this, a series of remnant plains each studded with higher remnant peaks or tablelands will be formed. Since such a process would be very slow in a hyperarid climate with only occasional cloudbursts, we probably have to look back to a time when the climate of Egypt was much wetter—the Pleistocene or even much earlier—to have produced the cycles of downcutting just described. In the deserts today, the process continues slowly 'nibbling' away at the residual higher peaks.

Recently the formation of the Nubian pediplain landscape has been linked to a proposal for a river system that preceded the Nile called the Qena River. This was a south-flowing river that existed during the Miocene from about 24 to 6 million years ago. This new proposal asserts that repeated uplifts of the region in connection with the opening of the Red Sea, rather than drops in base or sea level, maintained the elevation differential that led to the cycles of downcutting. (The Qena River is discussed in more detail in the chapter on the Western Desert.)

Several of the wadis whose mouths now form bays along Lake Nasser have interesting histories. The Kalabsha Wadi extends westward for nearly 100 km; today the lake waters penetrate about 10 km into this depression. In 1981, a magnitude 5.3 (on the Richter Scale) earthquake centered on this area raised fears that the weight of the lake's water was the cause. A study showed that the quake was associated with a previously existing east–west fault. The study concluded, however, that the Aswan High Dam could withstand any disturbance that was likely to occur.

Wadi Toshka (or Tushka) is 34 km northeast of Abu Simbel. It assumed a new importance when it was decided to construct a spillway through it to divert excess lake water rather than releasing it at Aswan and letting it sweep downstream. Archaeological and geological surveys were conducted in preparation for digging the channel (see Haynes, 1980). It was determined that this wadi had once carried water from lakes in the Kiseiba-Dungul

Depression to the Nile. Within the wadi the investigators found a channel incised into bedrock but now filled with sediment and covered with wind blown sand. The extreme depth of the channel indicated it had probably been carved during the late-Miocene Mediterranean desiccation, when the Nile was cutting its own canyon. Primitive flint implements among the upper layers of sediments indicated that humans might have lived along this stream as much as 200,000 year ago.

In the 1970s, the Sadat Canal was excavated through the sediments at the west end of the wadi to open a route for excess water to flow into the Kiseiba-Dungul Depression. In September 1996, when Lake Nasser's water level reached 178 m above sea level during an unusually high flood season, overflow water entered the depressions at the west end of the canal and began to form several lakes. Several more years of higher than average floods followed, so that by the year 2000 the Toshka Lakes occupied around 1600 sq km and held 20 billion cubic meters of water.

Wadi Toshka may have been an important route to a hard stone quarry about 70 km northwest of Abu Simbel. The quarry lies in an outcropping of the basement complex rock and was worked from Predynastic times to the Middle Kingdom. It provided the attractive anorthosite **gneiss** that was used for statues such as the famous Chephren (Khafre) with the falcon on his chair in the Egyptian Museum. During the Middle Kingdom, a road consisting of a cleared pathway was constructed for a distance of 80 kilometers to the Nile Valley at Toshka.

In 1997, construction of the Sheikh Zayed Canal was begun, leading from a point on the west bank of Lake Nasser just north of Wadi Toshka. This canal is designed to convey 5 billion cubic meters of water a year to the New Valley Project (now also known as the Toshka Project) in the depressions of the Western Desert. The water for the canal will have to be pumped up an average of 21 to 53 m to get across the intervening section of the Nubian Plateau.

Wadi el-Allaqi enters Lake Nasser from the east 125 km south of Aswan. The wadi's main channel is over 350 km in length. This channel and its major tributary, Wadi Gabgaba, along with hundreds of lesser tributaries, drain an area of the Nubian Desert estimated at 44,000 sq km. These must have been a major source of the Nile's water during earlier periods of wetter climates before the connection to the sub-Saharan sources (White and Blue Niles) was established. Even in these hyperarid times, an occasional thunderstorm in the region will convert the wadi into a raging torrent. In 1830, Linant de

Bellefonds, a Frenchman employed by Muhammad Ali, reported that a flood from the Wadi el-Allaqi was so strong that it prevented his dahabeya from sailing past the confluence of the wadi and the Nile. Historically, Wadi el-Allaqi was important to Egyptian society as the location of several productive gold mines. One of these at Umm Qareiyat continued to be mined into the 1900s.

Antiquities along Lake Nasser

Lake Nasser has obliterated the original Nile Valley south of Aswan. Our impression of the valley's former appearance must be gained from the archaeological reports of those who conducted salvage expeditions or from the far more colorful journals of nineteenth- and early twentieth-century travelers. Once south of the Aswan cataract, boats formerly cruised between high sandstone cliffs that rose several hundred feet above the river. At Kalabsha, 60 km south of Aswan, granite cliffs appeared and constricted the gorge to a width of only 200 m for a distance of nearly 5 km. Elsewhere, narrow floodplains supported small villages and some agriculture. Travelers such as Amelia Edwards (1888) described the lively village activities, the appearance of the monuments choked by sand, and the desolate expanses of desert. At Wadi Halfa on the Egypt–Sudan border and 345 km south of Aswan, the Second Cataract began and continued for nearly 200 km. Most travelers ventured no further. All of this picturesque scenery and many archaeological sites now lie beneath the waters of Lake Nasser.

In fact, the dams built at Aswan in the early decades of the twentieth century began the obliteration of Nubia. As the water impounded by these dams inundated riverbanks upstream, the Nubian villagers were forced to move their homes to higher levels or move away. When the High Dam was completed, more than 60,000 Egyptian Nubians lost their homes. They were resettled in new villages constructed for them east of Kom Ombo (see Chapter 7). Many were depressed at first by loss of their traditional homes, riverfront way of life, and the failure of some government promises to materialize. Gradually the resettled Nubians have been more content, as farming opportunities have improved. While the younger generation is becoming accustomed to the location, some of the elderly would like to move back to the riverside.

Some archaeological salvage work preceded the 1912 and 1934 phases of dam building, in order to excavate and record sites to be destroyed. In the late 1950s and early 1960s, an intensive international effort was mounted to record as much as possible about sites—many heretofore unknown—that would be submerged by the expanded lake. Like the Nubian people, some monuments were saved from drowning by relocation. Since only a small number could be preserved, it must have been an agonizing choice to decide which ones to save. And once chosen, the fates of many were left hanging while a worldwide fundraising effort occurred. The most incredible effort involved the relocation of the two rock-cut temples at Abu Simbel. The project involved engineers, stonecutters, and archaeologists from around the world. The $40 million cost was covered predominantly by Egypt and the United States and by other funds raised by UNESCO.

The Abu Simbel temples, built by Ramesses II to honor himself and his wife, were carved into a high cliff of Nubian sandstone on the west bank of the Nile about 280 km south of Aswan. This site is now submerged beneath the waters of Lake Nasser. To preserve the monuments, they were moved to a new site 65 m above the old one and 210 m to the west. The original temple façade was carved into a sandstone escarpment, while the interior was hollowed out of the cliff. The Nubian sandstone has many cracks and fissures in it, the sand grains being only weakly cemented together. While this made the work of carving the statues and reliefs easier originally, it meant that they were easily worn. If the temples had been placed anywhere but in an arid environment like Nubia, they would have been destroyed long ago. In the salvage operation, the fragile rock was consolidated in many places prior to moving it.

The temples were relocated by first removing the overburden, that is, the majority of natural rock above the temple, leaving only a thickness sufficient to serve as a 'roof.' Then the temples were cut into blocks that were removed to a storage area and later reassembled. The rebuilt Great Temple is covered by a massive concrete dome 60 m wide by 27 m high. Built to sustain loads of 55 tons per sq m, it is itself an engineering marvel that prevents the restored overburden from resting directly on the temple interior. Work on the move began in 1963 and was completed in 1968. In their new site, the temples retain their original orientation, so that at sunrise on October 22 and February 22 the sunlight still shines directly down the main aisle of the temple and illuminates the sanctuary with its four gods: Re-Harakhte, Ptah, Amun-Re, and Ramesses himself.

Other temples, tombs, and buildings from sites along the riverbanks were also relocated. Being smaller it was possible to move them longer distances and group several together in one place to create a series of destinations for travelers cruising between Abu Simbel and Aswan. For example, a large temple with sections dating from Amenhotep II to the Romans once stood on the riverbank at Kalabsha. It was moved to a promontory just southwest of the High Dam. This relocation process was facilitated by the fact that the temple was built of 16,000 quarried sandstone blocks and stood in the open rather than being cut into the cliff face. The success of this move provided an example during later discussions about the future of the monuments at Philae (see the next chapter). The blocks for the Kalabasha Temple, like the other temples and tombs of Nubia, were obtained at nearby sites, but these ancient quarries have also been inundated and so are not available for study. When one visits the relocated monuments, one is immediately struck by the deep reddish brown color of the sandstone in the structures and in the surrounding hills.

Aswan

Throughout history the River Nile was Egypt's main highway; conveniently one could float or row northward with the current, or sail southward using the predominant north winds. Aswan was considered the southern frontier of Egypt, since through-navigation was blocked by the **cataracts** or rapids that began just to the south. From garrisons situated in Aswan, military missions could be launched to subdue the Nubians. Trading missions into central Africa left from and returned to this point and then sent their trade goods of ivory, animal skins, ostrich feathers, ebony, and gold by ship downriver to Memphis or Thebes.

To make it possible for his ships to sail beyond Aswan, King Merenre of the Sixth Dynasty ordered a channel to be excavated through the rapids. The official he sent to supervise the task left a record of the work in his tomb, while the king himself was so pleased that he visited the site and left two inscriptions commemorating his visit on rocks in the river. The channel had to be cleared anew in the reign of Senusert III of the Twelfth Dynasty before he could undertake his conquest of Nubia. In the New Kingdom, Tuthmosis III again had to order work on the channels. It is unlikely that any of these clearances involved actual excavation of the granite bedrock, since technology was lacking to do this. Instead the goal was to remove some of the boulders and sediments brought down by the annual floods.

Travelers in the nineteenth century found the trip through the cataract with their rented houseboats, or *dahabeyas,* a memorable experience. Boats had to be pulled or winched upstream by many hands while the downstream ride between the granite islands through narrow channels filled with rushing water was hair-raising.

The geological setting of Aswan is complex. Igneous and metamorphic

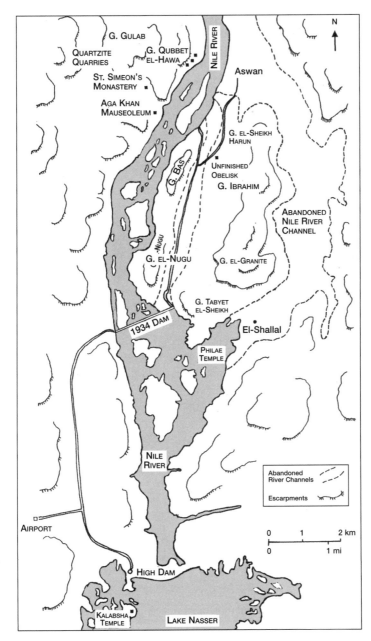

G. Gulab

Quartzite Quarries

G. Qubbet el-Hawa

St. Simeon's Monastery

Aga Khan Mauseoleum

Nile River

Aswan

N

G. el-Sheikh Harun

Unfinished Obelisk

G. Bas

G. Ibrahim

Abandoned Nile River Channel

G.-Nugu

G. el-Nugu

G. el-Granite

G. Tabyet el-Sheikh

1934 Dam

El-Shallal

Philae Temple

Nile River

Abandoned River Channels

Escarpments

Airport

0   1   2 km

0   1 mi

High Dam

Kalabsha Temple

Lake Nasser

Fig. 6.1:
Map of Aswan,
showing
the location
of dams and
quarries.
G = Gebel.

rocks of the deep-lying **basement complex** have been uplifted and exposed in this area. The river has eroded the overlying Nubian sandstone and carved deep channels into the igneous rocks. In addition to the island-choked channel of the present Nile, there is evidence for at least two other ancient, but now abandoned, Nile channels. The road leading northward from the east end of the 1934 Dam follows one channel through a narrow valley between Gebel Bas and Gebel Ibrahim. Another channel lies to the east of Gebel Ibrahim, where one of the earlier, more vigorous Niles carved a vast valley. This was probably the course of the late-Miocene Eonile, since a deep gorge has been detected under the Pleistocene sediments that fill it. This valley is the site of much new residential and commercial development. The steel mill, fertilizer factory, and other industries made possible by the High Dam's electric power output are located here. Most tourists are not aware of this thriving, but not very picturesque complex. The railroad from the north runs through the valley to the town of El-Shallal, which was formerly the port for river steamers from Aswan to Wadi Halfa. A branch of the railroad has now been extended to the new port on the shore of Lake Nasser beyond the High Dam.

Aswan's importance for the ancient Egyptians arose not only from its strategic geographic location, but also from the variety of stone that could be obtained there and transferred easily to barges on the river. Sandstone forms the western cliffs north of the 1934 Dam and survives atop the flat-topped hills of Gebel Bas and Gebel Ibrahim. The reddish brown color of the local sandstone is the result of the high concentration of iron compounds that oxidize on their surface. Northeast of Aswan are deposits that provide iron ore for the steel mills at Aswan and at Helwan south of Cairo. This source of iron ore may have been exploited as early as the New Kingdom, when the Egyptians learned about iron working from the Levant. Prior to that they may have mined the red **hematite** for pigments.

The softness of the sandstone in the escarpment of Gebel Qubbet el-Hawa on the west bank opposite Aswan encouraged governors of Aswan during the Old and Middle Kingdoms to build their rock-cut tombs there. Some of the sandstone on top of the cliffs has been altered into an extremely hard **quartzite**. This was quarried in pharaonic times to provide material for statues and sarcophagi—several examples of which can be viewed at Luxor. For example, the carved sarcophagus in the tomb of Tutankhamun is made of dark red quartzite, although it has not been tested to determine the quarry from which it came.

A large variety of igneous and metamorphic rocks is found among the numerous outcroppings of basement complex around Aswan, but only a few kinds of rock were exploited extensively in ancient times. Red or pink granite and gray granodiorite are the most prominent of these. Nearly one hundred granite-quarrying sites have been identified on the east bank and the islands. Many natural granite boulders still lie in heaps along the roads and riverbanks.

Granite was a popular stone among the Old Kingdom monarchs. It was used to construct the burial chamber at the Saqqara Step Pyramid in the Third Dynasty. Fourth Dynasty kings used it lavishly in their burial chambers and in their mortuary temples at Giza. Kings Djedefre, Khafre, and Menkaure had one or more courses of granite casing on the outside of their pyramids. Granite was also used for sarcophagi and portcullises to seal the pyramid passageways. During the Fifth Dynasty, granite portcullis chambers continued to be built in limestone pyramids. It has been estimated that 45,000 cubic meters of granite were shipped from Aswan to the Memphis area in the Old Kingdom alone.

Granite was surely preferred over other stone for some of these applications because of its greater hardness. In addition, its appearance may have been a consideration. Both red granite and black or gray granodiorite were used; they take a wonderful polish and the colors may have had a special symbolism in certain applications. For whatever reason it was chosen over other stones, granite was much more difficult to work than either of the other far more common building materials, limestone and sandstone.

Much nonsense has been written about how the ancient Egyptians must have had advanced techniques for working hard stones and how knowledge of these techniques has since been lost. There is no evidence for any unknown methods and much evidence for the methods actually used, which mostly involved brute strength.

Most of the granite employed during the Old Kingdom was probably not quarried in the sense of extracting a block from a great mass of mother rock. Instead naturally formed blocks were used; all that was necessary was to select a block of appropriate size, load it onto a barge for shipment downriver, and finally shape it. Judging from unfinished examples left in the quarries, preliminary shaping was done at Aswan to reduce the weight.

Granite naturally weathers into conveniently-shaped boulders (Fig. 6.2). Granite forms from molten magma as a great solid mass far underground. As this mass cools and solidifies it is typically under great pressure. Tectonic movements may eventually lift this mass closer to the surface. As the land above this granite mass erodes, the external pressure on the mass is reduced and the granite begins to fracture along horizontal and vertical planes at right angles—an action known as **pressure release**. This produces **joints** that define nearly perfectly rectangular blocks. Tectonic forces can also crush and fracture the rocks. Slightly below ground, in a humid environment such as Egypt experienced in times past, the blocks weather along the joints into more rounded shapes. Then when the rocks are exposed by erosion of the overburden, the blocks are revealed ready for use without the difficulties attendant to quarrying them. The role of the river is primarily to expose the rocks, but some rounding may also occur when the floodwaters sweep debris over the surface of the boulders.

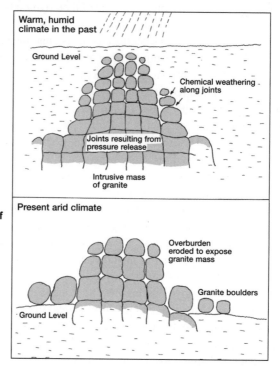

Fig. 6.2:
The natural weathering of an intrusive mass of granite into rounded boulders.

After selecting a granite boulder of appropriate size and shape, the Egyptians used a number of methods to shape and smooth it. These included bashing the surfaces with **dolerite** hammer stones, abrading with sand (which, containing quartz, can cut the quartz and other mineral grains in the granite), and sawing or drilling using copper saws and drill bits in conjunction with fine sand. All of these would be slow, time-consuming operations, but each would eventually produce the desired results. An Old Kingdom sarcophagus, whose lid cracked after it had been partially sawn off the bottom trough, can be observed in the Egyptian Museum (Room 47, lower floor). With the aid of a flashlight, one can clearly see the saw marks in this hard stone.

The already-jointed granite blocks probably served many of the purposes of the Old Kingdom, such as casing blocks and wall blocks in chambers and temples; even pillars and roofing beams could be made from larger stones. But when demands exceeded the size of the available boulders, quarrying became necessary. Thus the obelisks and colossal royal statues popular in the Middle and New Kingdoms had to be derived from a larger, solid mass of rock.

Evidence of the methods the Egyptians used to do this can still be seen on the famous 'unfinished obelisk' in the Aswan quarries. If completed, this massive monument would have been almost 42 m tall and weighed 1,168 tons. Along each side of the obelisk are trenches formed by overlapping shallow depressions—each one revealing where a worker stood and hammered on the granite with a dolerite pounder, a ball of stone weighing 4 or 5 kg (around 10 pounds). The reason that this obelisk was abandoned after months of work is obvious: an enormous crack stretches across the top.

Surely you will wonder: what caused this crack, and why didn't they notice it sooner? I believe the crack occurred suddenly, as a result of pressure release. The future obelisk had still been under compression from the surrounding rock before the shaping began. As the trenches were formed to define the sides, the stresses within the obelisk's shaft were no longer balanced by the surrounding rock, and they 'relaxed' into the crack we see today. One can only imagine the feelings of the workers who had labored so long and the supervisors who had a deadline to meet.

In the Aswan quarries, and on many sites throughout Egypt where granite was employed anciently, one can see granite blocks with a series of rectangular holes defining the line along which the block is to be split. It is sometimes asserted that wooden wedges were inserted into these holes and soaked with water until they swelled and split the rock. Recent experiments with wetted wedges show that they would not have been effective in splitting granite. The holes themselves must have been cut using iron or steel chisels, which only became available in Egypt in the Late Period and widespread in Greco-Roman times. In these classical times, iron wedges and possibly splints were used after a series of rectangular holes had been cut. Where employed, two splints (also called fins or feathers) in the form of two flat metal plates were placed into the holes, and the wedge was placed between them. The splints served to direct the force of the wedge horizontally against the sides of the holes rather than allowing it to be driven into the rock at the bottom of the hole. These methods of rock splitting are still in use today. Many of the holes seen in granite blocks on archaeological sites may date from an even more recent period and be the work of modern stone robbers.

Once a granite block had been quarried, it was moved to a dock on the Nile along a series of specially constructed roadways and transferred to a boat to be floated down the Nile to a building site such as Giza. Travelers visiting Giza can look at the way in which these granite blocks were incorporated into buildings. When granite was used for casing blocks on a pyramid or to line the walls of a temple, only the block's sides, top and bottom, and front were dressed. On the backside, the softer limestone backing-stones were cut to accept the natural curves of the granite blocks. Examples of this can be seen on Khafre's pyramid, where several casing blocks remain in place, and in the Khafre valley temple and Menkaure mortuary temple, although most of the granite originally placed in those temples has since been stripped away. We can also see that the casing blocks on Menkaure's pyramid have been set in place with their edges reduced to the final face angle, but excess material (in fact the natural curve of the block) left in the centers.

In the first half of the twentieth century, Aswan granite was used to build the series of Aswan dams designed to store part of the floodwater for summer irrigation. The water level in the lake behind the dam rose in the late fall as the last stages of the floodwater were retained, then it dropped in the summer. The island of Philae was just upstream of the dam, and the splendid sandstone temples on it were submerged for nine months each year. In the early 1900s, methods of protecting the monuments were discussed. Removal to another site was considered, but it was felt this would destroy the essential context of the temples, so they were left in place. The fragile sandstone was reinforced, however, which turned out to be a crucial decision, since this protected the blocks against the erosive action of the water for almost seventy years. The plans for the High Dam called for the sluice gates in the 1934 dam to serve as regulators of the discharge into the river. As a consequence, the island of Philae would be permanently submerged. Again the question of how to preserve Philae's monuments was raised, and the plan of building a coffer dam around the island to hold out water was adopted first. It soon became clear, however, that such a dam could not be relied on and that relocation was the best solution. Nearby Agilka Island was selected as the new site; its granite peak was excavated and rubble dumped around its edges to create a topography that more closely resembled the original Philae. The more than 37,000 individual quarried blocks that made up the many different monuments were carefully recorded, then they were disassembled, cleaned of seventy years' worth of silt and organic incrustations, and reassembled in their new home. Work began in September 1975; it took eighteen months to dismantle the monuments and another two years to rebuild them.

When the High Dam was being designed, it was expected that the granite outcropping above the First Cataract and in the cliffs would form an excellent impermeable substrate for the massive dam. But when engineers bored test holes into the river bottom, they discovered a huge sediment-filled gorge extending more than 250 m below the modern Nile channel. In the early 1960s no one could explain how such a deep canyon had been formed. When the late-Miocene desiccation of the Mediterranean was proposed in 1970, the mystery of the Nile Canyon was solved.

In order to accommodate itself to the site, the High Dam was designed as a massive rockfill gravity dam structure. It is 980 m (or almost a kilometer) thick at its base and rises 111 m from the riverbed to a crest 40 m wide. Its volume is equivalent to seventeen Great Pyramids. In fact, the north–south cross-section of the dam resembles a pyramid, with two wings extending upstream and downstream; these wings incorporate the cofferdams that were built on each side of the construction site. To prevent water from percolating through the sediments filling the gorge beneath the channel, a 'grout curtain' was formed by pumping cement into deep holes bored to bedrock in a line across the entire channel. The pumping pressure caused the cement to spread out and consolidate the surrounding gravels, sands, and silts. This grout curtain is 60 m thick and extends about 800 m from the dam's core down to the granite bedrock. A fascinating account of the dam's construction can be found in Little (1965). The feasibility of the construction plan depended on the availability of local materials (granite and sand) and a source of electric power at Aswan. The 1934 Aswan Dam did not include a hydroelectric power plant when it was first built, but in 1961 one was constructed in it specifically to provide power for the High Dam construction project.

The Nile Valley from Aswan to Luxor

L eaving Aswan to cruise northward on the Nile, we see that cliffs of Nubian sandstone bound the river closely on both sides. The valley varies in width, with narrow stretches of floodplain and cultivation appearing on one side or the other. The tops of the sandstone cliffs form a wide pediplain—a plain whose formation dates to the late Pliocene. Layers of rocks younger than the Nubian sandstone once overlaid this plain, but they were removed by erosion. The sandstone plain is bounded on its west side by a high escarpment of Eocene limestone that is the eastern edge of a great limestone plateau. At the base of the escarpment—between the Eocene limestone and the sandstone—layers of shale and chalk of late Cretaceous or Paleocene ages are exposed at various points. Perched on the sandstone pediplain are isolated hills of Eocene limestone surrounded by shales. Examples of such hills include Gebel el-Barqa, Gebel Rakhamiya, Gebel el-Nazzi, and Gebel el-Surai. Some of these hills represent massive blocks that slipped from the more distant limestone cliffs during the late Miocene period while the Nile Canyon was being formed.

On the east side of the river, the sandstone plain slopes upward to the Red Sea Mountains. In these mountains are igneous and metamorphic rocks of the ancient basement complex that have been uplifted and denuded of younger rocks. The map of Figure 7.1 shows these features as well as a number of the major wadis originating in the Red Sea Mountains. The point where a wadi met the floodplain was often chosen as a town site, since the wadi served as a convenient path into the mountains for miners and traders. When they returned to the river, they often transferred their goods from caravans to boats to continue their journey.

During the Pliocene, the Nile Canyon was partially filled by marine

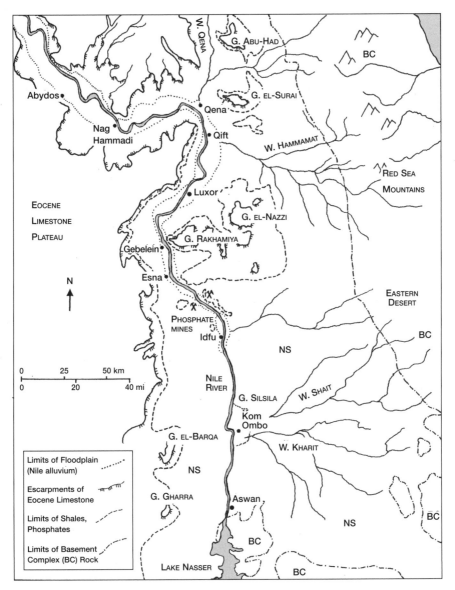

Fig. 7.1: Map of the Nile Valley from Aswan to Luxor.
G = Gebel, W = Wadi, NS = Nubian sandstone exposed,
BC = basement complex exposed.

deposits that were then overlaid by riverine deposits carried by the Paleonile. These deposits completely filled the canyon to a height of at least 180 m above current sea level—perhaps because the world sea level of that era was that much higher. As sea levels fell, a succession of Niles began to cut their channels into this filling and the sandstone plains to either side. Geologists studying the area between Aswan and Idfu found evidence for a series of channels cut into the sandstone bedrock at levels of 180, 150, 160, and 130 m above present sea level. These abandoned river channels lie to the west of the current river's course. Each channel is marked by a belt of river deposits topped by a layer of hard, cemented gravel. Evidently the gravel was laid down when each successive river waned and lost the ability to transport its larger sediments. Later, when rainfall increased again and permitted a new river to flow, that river found it easier to cut a new channel in the soft sandstone bedrock than to carve through the deposits of the earlier channels. Geologists dated all of these channels to the late Pliocene to mid-Pleistocene. When the modern River Nile finally began to flow, it chose a new, more easterly course and cut a channel to around 80 m above current sea level. It seems to have followed the course of the ancient Miocene Nile Canyon.

Kom Ombo

About 35 km north of Aswan, the sandstone cliffs fade away into the distance and a vast plain appears. This plain has an area of 900 sq km and an elevation 15 m above the flood level of the modern Nile, so this is not a floodplain produced by the modern river. The town of Kom Ombo sits on this plain on the east bank of the river. Most tourists on cruise ships will stop long enough to visit the Greco-Roman temple dedicated to Sobek and Haroeris, but they will not have a chance to look farther afield at some very interesting geological features that may be more visible to those traveling through this region by train. These features were important in the ancient and more recent history of this area.

The Kom Ombo plain is defined by faults to the north and south, where the bedrock is uplifted relative to the plain. During the late Pleistocene this plain was the depository for huge quantities of sand and then gravel washed from the Red Sea Hills. Wadi Shait and Wadi Kharit, with their numerous tributaries draining a huge part of the Eastern Desert, carried these sediments and dropped them on the plain. These sediments were then covered by a deep layer of fertile silt carried by the Nile from the proposed Lake Sudd.

The excellent agricultural potential of the Kom Ombo plain was exploited in Ptolemaic times as a city there replaced Aswan as Egypt's southern frontier town. In the early 1900s, this plain became the focus of a great agricultural commercial venture financed by European investors. Water was pumped from the Nile, whose flow was augmented during the spring months by water impounded by the dam at Aswan. This perennial irrigation permitted the growth of cash crops such as sugar cane and cotton. In the 1960s when it became necessary to find a place to relocate the Nubians who would lose their homes to Lake Nasser, several new villages were built and land prepared for cultivation. Around 50,000 Nubians moved to these villages, whose names—such as New Abu Simbel and New Toshka—recall the original home sites of the inhabitants.

The Kom Ombo plain is bounded on the north by a cliff of Nubian sandstone that has been uplifted nearly 200 m. The river has carved a channel through the cliffs that is only 400 m wide. An older, now abandoned channel lies to the east; the Cairo–Aswan railroad passes through this depression. On the river's east bank is a mountainous mass of sandstone called Gebel Silsila. Building stone was quarried here during the New Kingdom and transported downriver to Luxor. This quarrying continued until the Greco-Roman period. The rock mass has well-spaced vertical joints that facilitated quarrying. Some rock-cut tombs were placed here, along with many stelae and graffiti. Some of these are visible from a cruise ship and may be visited.

From Idfu to Esna

Idfu (also spelled Edfu) is usually another stop on the cruise ship itinerary. Tourists visit the Ptolemaic temple also built of Gebel Silsila sandstone and then reboard their ship for lunch. They are thus apt to miss an important geological transition. Between Idfu and Esna the sandstone cliffs give way to limestone. The sandstone stratum continues northward, but it lies beneath the more recent limestone layers. The entire 'rock sandwich' tilts toward the north, but at a steeper angle than the river's gradient. Thus as one travels north, one sees younger and younger rocks facing the river. The limestones are not the only kind of rock in the cliffs: layers of other kinds alternate with them.

One of the rock layers revealed in some outcroppings in the stretch north of Idfu contains a high percentage of phosphate derived from ancient animal remains (bones and **coprolites**). Formed at the end of the Cretaceous, some 83 to 65 million years ago, this layer lies between the Nubian sandstones and the Dakhla shale. Phosphates are mined here, since they along with nitrates and potassium are the main ingredients in fertilizer. This phosphate layer actually occurs over a wide area of the country, but here it is exposed near the river at a location convenient for shipping down the Nile. The placement of nearly all mines and quarries can be understood in terms of two factors: the quality of material available and the need for workers to reach the quarries and to transport away the materials removed from them.

Travelers will be amazed to find that the sandstone temple at Esna is nearly 9 m below the current ground level. Over the centuries, occupational debris and silt deposited during the annual inundation raised the floodplain at a rate of a few millimeters each year. The ground level is no longer rising, since the Nile silts are now trapped behind the High Dam. The loss of this fertile replenishment is not affecting agriculture alone; it is also having an impact on home construction. From the earliest times until just a few years ago, houses in Egypt were built of sun-dried mud bricks. These bricks were made of Nile mud mixed with straw or dung to prevent the bricks from cracking as they dried. Mud served as mortar to build walls, and mud plaster was applied over the outer surface to protect the bricks from the infrequent rains. Eventually such houses deteriorated, but they were easy to replace with local materials.

After the completion of the High Dam, brick makers could no longer rely on an annual mud supply and began to buy topsoil from farmers' fields. So much agricultural land was adversely affected that the government banned the practice. Since then, brick makers have been searching the desert for sources of clay for bricks. Clay bricks are then baked using the ample supplies of petroleum discovered in Egyptian offshore waters. Crushed basalt or granite can be added to the bricks to prevent them from cracking during firing. Baked bricks are used to build house walls, after columns and floors are formed of concrete. Because bricks are heavy, but low in value, they are typically made locally to avoid the transportation costs. Today each governorate has its own brick factory. The use of lime cement in mortar and concrete is also a recent innovation. Lacking fuel to transform limestone into lime—a process requiring high temperatures—all ancient construction relied on mud mortar and **gypsum** plaster. Today, as one travels through towns and cities, one can see the homes built in the newer and more permanent style using poured concrete floors and columns and baked brick walls alongside the older models of mud brick. What has not changed is the practice of using the roof as a storage area for fodder, fuel, and extra supplies. Many houses appear unfinished, with steel reinforcing rods sticking up out of bare concrete columns, but these upper areas will eventually be finished to house a son and his family.

As we continue our journey toward Luxor, the floodplain becomes wider. This plain is the result of a recent aggradation phase in the Nile's history. In the early Pleistocene, the river—responding to falling sea levels—carved a series of channels deeper and deeper into the Pliocene filling of the Nile Canyon. This created a series of terraces at successively lower levels along the sides of the valley. Toward the middle of the Pleistocene Epoch, humans began to inhabit such riverside terraces, and remains of their flint implements have been recovered, helping scientists to date the terraces. At some point the river channel reached its lowest point, and deposition of sediments, in the form of fine silts, began to refill the valley. The river may have actually experienced several rounds of degradation and aggradation, but evidence for this is now hidden beneath the recent alluvium.

Luxor

The city of Luxor (called Thebes by the Greeks, Waset by the ancient Egyptians) rose to prominence in the Middle Kingdom. Many of the outstanding temples were started during its peak of prosperity in the Eighteenth Dynasty and were added to in later periods. Most of these temples, like the ones in Idfu and Esna, were built of sandstone quarried at Gebel Silsila. Building with sandstone was a Middle and New Kingdom development—Old Kingdom temples and tombs were primarily built from limestone. The reason for the change is entirely explained by availability of the two kinds of stone. Most Old Kingdom tombs and temples lay near the northern city of Memphis—a region in which limestone is the prevailing bedrock. At Luxor, the nearest limestone cliffs lie 4 km to the west and nearly 20 km to the southeast of the city and the river: too far to transport enough blocks for a temple overland. One outcropping of Eocene limestone at Gebelein (on the west bank of the Nile 40 km south of Luxor) was exploited, but it may have been inadequate for the extensive building needs in the New Kingdom. By contrast, the huge sandstone quarry at Gebel Silsila was next to the river, so blocks could easily be transported downstream to the docks at Luxor. We notice that the major pharaonic monuments on the east bank lie close to the river: blocks did not have to be dragged far inland. In fact, a canal ran from the river to a quay in front of each temple in ancient times.

The construction of these New Kingdom temples did not present the same challenges to their builders as the Fourth Dynasty pyramids did to theirs (which we will discuss in Chapter 10). The New Kingdom structures required less material, generally used smaller blocks of stone, did not raise stone blocks to such great heights, and did not need to support the gigantic weight of a pyramid over open internal spaces. Each New Kingdom temple had the same basic components proceeding outward from a central sanctuary: hypostyle halls, colonnaded courtyards, and pylons. Karnak Temple's massive plan, for example, was achieved by replicating these components over and over again with additions and alterations made by successive kings. Despite the complexity of its plan, the building methods at Karnak, like those in most other temples, were very straightforward and employed a post-and-beam form of construction (Fig. 8.1). Vertical columns (the posts) supported horizontal members called **architraves** (the beams). These architraves were made up of one or more beams rectangular in cross-section and set with their longer edge vertical for added strength. A layer of roofing beams was placed

Fig. 8.1: Diagram illustrating the post-and-beam construction of the Colonnade Hall in Luxor Temple.

at right angles over these architraves. Although some temples had chapels or other rooms on their roofs, the columns and architraves supported relatively little beyond their own weight. The sandstone columns were built up of sections (drums) of whole or half cylinders. This method was much easier than cutting, transporting, and erecting enormous monolithic columns.

Several authors have said that sandstone was employed in New Kingdom temples in preference to limestone because longer beams could be cut from the sandstone. The implication is left that sandstone, in general, is stronger than limestone. This is not true. Both sandstones and limestones can vary greatly in their strength when used as a beam, depending on their chemical composition, the texture of the grains, and the amount of cement between the particles. Some limestones are much stronger than many types of sandstone. What is probably true is that there was a difference in the condition of the material in the quarries from which stone for the Luxor temples could be obtained. Limestone is very brittle, and in most quarries the material is excessively jointed. This makes it difficult to cut long or large blocks of stone. The sandstone at Gebel Silsila is jointed, but not excessively so, hence long blocks could be extracted, making it possible to space huge columns at considerable distances apart and still bridge the space between them.

Nubian sandstone is easy to cut and carve. The deeply-incised hieroglyphs, thought by some to have been ordered to prevent usurpation or defacement by later monarchs, are very effective on exterior wall surfaces, where the play of light over the carvings accentuates their details. Like all other carved areas these walls would have been brightly painted. We know that mud-brick ramps and a filling of mud or rubble were built up course-by-course to permit the moderate-sized blocks to be hauled up and set into place. Finishing of the stone surfaces and decorating of the walls and columns took place from the top down as the mud fill was removed from the interior. Scaffolding was probably used to access the exterior walls for carving and painting.

Another building material, unfired mud brick, was used in the Luxor temple complexes to build the gigantic enclosure walls. Remains of such a wall can be seen around the Karnak Temple, although an even better view of one may be obtained at the Temple of Hathor in Dandara. Mud bricks were also used to build extensive storage magazines for temple supplies and religious equipment. A series of these magazines can be seen around the main temple of the Ramesseum on the West Bank. Some visitors may be surprised to see that they have vaulted roofs and wonder if they can be contemporaneous with the temple built in the Nineteenth Dynasty. In fact the ancient Egyptians were building these interesting **leaning vault** brick roofs as early as the Third Dynasty. They used them, as here, to cover narrow chambers or corridors. If one looks at the solid back wall of a chamber, one can see how the first few courses of bricks in the vault were set leaning against the vertical back wall. The first course had only one or two bricks in it, the next course a few more, and so forth. Subsequent courses leaned against the earlier ones and eventually reached high enough to complete the arch over the top. This building method had the advantage of not requiring any support (called centering) beneath the vault while it was being built.

The Open Air Museum at the Karnak Temple is an architectural and geological treat that is easily overlooked, or bypassed by the rushed traveler. It is entered via a gate in the north wall of the first courtyard inside the temple but requires a separate admission ticket to be purchased at the ticket booth outside the temple. Several small monuments have been reconstructed here whose components were discovered, reused or discarded, during restoration on other parts of the temple.

The stones of three barque stations were found. These simple structures were built to temporarily house the barque (boat) on which the shrine of the king or a god was carried in procession during a king's Heb Sed or festival of renewal. One was built for Senusert I of the Twelfth Dynasty. It is formed of a pure white limestone and is carved inside and out with scenes relating to the ceremonies. This illustrates very well the use of this fine-grained stone for delicate relief carving. Two other chapels were built of **Egyptian alabaster**—one for Amenhotep I, and one for Tuthmosis IV. The stone in these shrines is actually a form of calcium carbonate or calcite; geologically it would be termed travertine. The shades of swirly brown make this stone very attractive. It is a soft rock and lends itself to the production of very detailed carvings and inscriptions.

Blocks of another dismantled structure were rescued from the Third Pylon, where Amenhotep III had used them as filling. They have been used to reconstruct the Red Chapel of Hatshepsut. The name derives from the deep red color of the quartzite blocks forming the walls. Black granodiorite was employed to form the dado along the base, the cavetto cornice at the top, and the doorjambs. The restorers (from the Franco-Egyptian Center for the Studies of the Karnak Temples) used tinted masonry to fill gaps in the walls and uncarved blocks of the same stones to complete the structure.

Throughout the temples on the east bank are other examples of exotic stone. For example, the traveler can admire several granite obelisks and colossi that began their journey in the quarries at Aswan. They show the ancient artist's mastery in shaping and carving these hard stones. Reading about the difficulties that modern engineers have had in moving various obelisks from Egypt to Europe and America, one gains added respect for the ancient Egyptians' ability to transport and erect such gigantic monoliths. Much has been written about the possible methods employed, and the traveler is urged to read more if interested.

Some monuments were damaged in ancient times and bear evidence of ancient attempts to restore them. A statue of Ramesses II at the entrance to the Luxor Temple has a large crack in it; spanning the crack is a hole to accept a dovetail cramp.

Threats to Ancient Monuments from the Rising Water Table

In discussing the consequences of the Aswan High Dam in a previous chapter, I noted that the water level in the River Nile that used to fluctuate by about 6 m over the course of the year from high flood to low water now lies between the former low water and the flood levels. Formerly, fields were inundated for 40 to 60 days during the flood season, then the water drained into the lowered river, carrying with it excess salts. Today, the higher water level in the river interferes with the gravity draining of the irrigation water. This problem is exacerbated by the fact that more water is now being applied to the fields under the practice of perennial irrigation than when basin irrigation was employed and the fact that agricultural drainage and domestic sewer systems in Luxor are generally inadequate. The inevitable consequence is that the ground is becoming saturated with increasingly salty water. The sandstone temples are being gravely affected by water that is being drawn up into their columns and walls. One can see the height to which the water has risen by the chalky white discoloration of the sandstone blocks.

Along with the water, dissolved salts are carried from the soil into the stone. When the water evaporates from the surface of the stone in the dry air, the salts remain behind and recrystallize on or just below the surface. If the crystals form on the surface they conceal the inscriptions, whereas if they form under or between the sand grains, grains are separated from the walls. In this way inscriptions over large areas have been completely obliterated.

74

Several Egyptology projects are devoted primarily to recording the inscriptions from temple walls before they disappear. Recently, some of these projects have added attempts to conserve the carvings to their mission rather than simply recording them. Efforts are also underway to shield the stone from the ground water by placing an impermeable layer beneath walls and columns. The ultimate solution of reducing water use and draining sewage and irrigation water is still in the planning stages, although the necessity for this has been recognized.

The West Bank

No first-time traveler will miss the photo opportunity offered by the Colossi of Memnon sitting next to the main road from the Nile to the West Bank necropolis. Almost 18 m tall and each weighing more than 700 tons, they are the most impressive remains of the Mortuary Temple for Amenhotep III that once lay directly west of them.

The statues were carved initially from single blocks of **quartzite**, a term used in Egypt for a form of sandstone in which the grains are cemented together with quartz. Whereas sandstone, especially Nubian sandstone, is soft and easy to carve, quartzite (or more properly siliceous sandstone) was one of the hardest stones used by the ancient Egyptians. It was a favorite material for statuary and sarcophagi. There are two places in Egypt where quartzite can be quarried: at Gebel el-Ahmar northeast of Cairo and at several sites on the west bank of the Nile at Aswan. Since the New Kingdom builders were bringing so much granite and granodiorite downstream to Luxor from Aswan, one might suspect that the quartzite for these two gigantic statues came from there as well.

An Eighteenth Dynasty inscription at Aswan shows Men, an "overseer of works in the Red Mountains" and "the chief of the sculptors in the great and mighty monuments of the king," making an offering to a seated statue of Amenhotep III that looks very much like the Theban Colossi (Habachi, 1965). This inscription seems to support an Aswan origin for the Colossi. On the other hand, an inscription commissioned by Amenhotep III's great overseer of works, Amenhotep the son of Hapu, states that he "brought great and huge monuments as statues of his Majesty in excellent work, taken from On of the North to On of the South" (Habachi, 1965). The back pillar of the southern Colossus also speaks of "On of the North," or the city of Heliopolis, near where the Gebel el-Ahmar quarry is located. Egyptologists differ in their interpretations of these inscriptions and thus on the most probable source of the quartzite in the Colossi of Memnon. Recently, some geologists have attempted to settle this debate.

The statues are not in their original condition, but have been restored. The accepted chronology for this is that they were damaged by an earthquake in 27 BCE and restored by orders of the Roman emperor, Septimius Severus, around 200 CE. Some authors have doubted that the damage was entirely caused by a single earthquake and have proposed thermal expansion or pressure release as other contributory mechanisms. Others have proposed deliberate vandalism using fire to damage the statues. It is thus highly likely that the monolithic rocks were already weathered and that an earthquake simply toppled part of an unstable formation. In any case, the repairs took the form of replacing the missing torso and head of the northern colossus with a series of smaller stones. The pedestals, which are separate pieces of stone 4 m high, were also repaired, with the back half of the northern one also being replaced with new stones.

Geologists took samples of quartzite from Gebel el-Ahmar and three quarries on the west bank at Aswan. They studied the rocks microscopically to determine their composition, the shape and size of the grains, and the kinds of cement. They also measured the types and amounts of minerals and trace elements present. They decided that they could distinguish rocks from the different quarries using these criteria. When they examined samples from the Memnon colossi with the same techniques, they concluded that the original blocks of stone used by Amenhotep III's artists for the statues and pedestals came from Gebel el-Ahmar near Cairo, despite the long, upstream journey that would entail. By contrast, all the blocks used in the third century CE restoration came from a quarry at Aswan.

It is only fair to tell you that not everyone accepted the results of these studies. Some other geologists have questioned the conclusions on the grounds that rocks in a quarry are too heterogeneous chemically to be characterized by only a few samples, especially when the rocks of greatest interest have been removed. Furthermore, another quartzite quarry has recently been located on the east bank at Aswan that was not included the study. Such methodological debates as these simply reflect good science and should not detract from the promise such studies hold for answering questions of archaeological interest by geological means.

Temples and Tombs

While the monuments on the East Bank are mostly temples, dominated by the gigantic temples of Karnak and Luxor, on the West Bank we find both temples and tombs. The earliest monument is located in the natural amphitheater at Deir el-Bahari, where there are the remains of at least three temples. The earliest is the southern one, built by Mentuhotep II of the Eleventh Dynasty after his reunification of Upper and Lower Egypt following the First Intermediate Period. The monument contains both a mortuary temple and Mentuhotep's burial place, along with the graves of his wives and daughters. This conjunction is reminiscent of the Old Kingdom pyramid complexes, with each king's mortuary temple affixed to the east face of his pyramid and his relatives' tombs dispersed around his own.

Later practice separated the mortuary temple and the burials and placed the latter in a less conspicuous location. The idea of building tombs underground or in a cliff face was not new in the Eighteenth Dynasty, when the first tombs were built in the Valley of the Kings. Egyptians had been doing that since early dynastic times. What was new was the idea of separating completely the king's burial site and his mortuary temple. The plan was to build a highly visible mortuary temple on the floodplain. Staffed by priests to perform the necessary offering services, and endowed in perpetuity, it would be safe from vandalism. The tomb and its treasures, by contrast, would be hidden in a secret location. For this purpose, the Valley of the Kings seemed ideal. Hatshepsut's father, Tuthmosis I was possibly the first to commission such a hidden tomb, constructed, as his architect was to report, "no one seeing, no one hearing." His reign was short, however, and it was left to his daughter to make an architectural statement.

Hatshepsut's Mortuary Temple, the northernmost of the three at Deir el-Bahari, is a beautifully restored architectural marvel in a dramatic setting. The temple's placement, design, and building material all echo those of Mentuhotep's. Curiously, both monuments are primarily constructed of limestone with occasional use of sandstone. The limestone probably came from a nearby quarry and is of better quality than most of that type of rock available in the vicinity.

The temple at Deir el-Bahari had its share of granite adornments as well: six enormous granite sphinxes were placed on the second terrace. Fragments of these—for they had been viciously destroyed, perhaps on orders from Tuthmosis III—were recovered and can be seen, now restored, in the Egyptian

Museum in Cairo and the Metropolitan Museum of Art in New York. Mere painted sandstone was used for the double row of sphinxes that lined the road from the temple to the cultivation. What survived Tuthmosis III was damaged by rock falls from the cliffs above. In fact, the adjacent temples of Mentuhotep and Tuthmosis III were almost completely destroyed in this manner.

Geology on the West Bank and in the Valley of the Kings

The limestone cliffs behind the Deir el-Bahari temples belong to what geologists have called the Serai **Formation** of the Thebes **Group**. This formation consists of a number of rock layers that alternate between dense limestone and softer layers of **marl**—a stone containing calcium carbonate mixed with clay. Marls are easily eroded and thus form sloping mounds of debris. By contrast, the limestone layers are more resistant and form sheer vertical faces. Overall, the limestone strata of the Serai Formation total 290 m in thickness, rising to the peak known as el-Qurn, visible slightly to the left and above the Hatshepsut temple. This natural pyramid may have inspired the Egyptian kings to select this site for their tombs.

The bottommost layer of the Serai Formation is a massive 120 m thick. It is the tan rock with vertical joints visible just above Hatshepsut's temple. Geologists refer to this thick layer as Member One within the Serai Formation, and we will follow their convention.

The Serai Formation rests on an older rock stratum called the Esna Shale. This is the grayish layer—apparently made up of many thin horizontal beds—directly behind and under Hatshepsut's temple. The shale–limestone contact is visible just above the stone retaining wall built over the third terrace. The Esna Shale layer is at least 60 m thick. It can be traced at the base of the cliffs as a sloping **scree** of material to the north and south of Hatshepsut's Temple. It is into these rock layers—Member One and the Esna Shale below it—that the ancient Egyptians excavated their royal tombs.

The names "Valley of the Kings" and "Valley of the Queens" are more impressive than Royal Riverbeds, but that is what each valley is—a giant wadi or dry riverbed. Under the current arid conditions they seldom carry water, but the occasional thunderstorm can dump tons of water into these channels in a very short time. This water, carrying loads of debris, can sweep into the tomb entrances and penetrate far into the down-sloping interiors. There it forms **alluvial fans**—dropping the larger pieces near the tomb entrance and carrying the finer material deeper.

As you enter the Valley of the Kings, you walk between the high cliffs made up of Member One of the Serai Formation. One does not have to be a practiced geologist to see that the rock is exceedingly fractured, with many open joints running vertically. Large nodules of flint—darker brown against the tan limestone—are also much in evidence. The ground at the base of the cliffs is littered with fallen rock fragments. Up until recently the valley floor was covered with piles of such rubble—products of rock falls, wadi wash, and ancient excavations. Recently, the Egyptian authorities and archaeological expeditions have been systematically removing all rubble to expose the bedrock in an effort to find additional tomb openings and artifacts concealed in the rubble.

It was not difficult for the ancient Egyptians to excavate into the limestone. Most of the coarse work was probably done using a flint pick hafted to a wooden handle. Once the general shape of the corridors and chambers was achieved with this tool, the walls could be finished using a copper adze. In the underground chambers two or more pillars of the bedrock would be left intact to support the flat ceiling. While the designs of the individual tombs vary, they usually have a series of stairways and corridors penetrating deep into the cliff, where the burial chamber is located; the stairs and slanting corridors also lead downward many meters. So a tomb excavation that began in the Member One limestone stratum could accidentally penetrate into the Esna shale below. This actually happened in three important tombs: KV7 (Ramesses II), KV17 (Seti I), and KV20 (Hatshepsut). The consequences of this will be described below.

The softness of the limestone made it easy to carve inscriptions, but the flint inclusions in the limestone would have made it difficult to achieve a smooth wall surface. Flints might leave a hollow, and areas of weak rock would crumble. So before the walls were decorated, they were covered with one or more layers of mud or gypsum plaster to smooth the surface. If reliefs were desired, the plaster layer could be built up to the requisite thickness for carving. Otherwise, painted scenes served as the sole decoration.

The scenes in many tombs survive in astonishing brilliance. This is because the ancient artists used mineral pigments in their paints, not vegetable dyes. Different iron oxide compounds produced reds, yellows, and browns; ground azurite and malachite gave blues and greens; powdered charcoal or soot provided black, while gypsum or calcium carbonate was used for white. The pigments used are still being investigated, and they turn out to be more diverse and complex than first thought—some even use an artificial substance (called **frit**) manufactured from a mixture of minerals. The loss of color that we observe on scenes is not the result of the pigments themselves fading, but of plaster peeling from the rock surface or of mechanical abrasion removing the paint layer. Mechanical abrasion can be caused by flood-borne debris or careless tourists leaning on walls.

The ancient engineers were not unaware of the dangers of flooding in the valley, and they made provisions to guard against it. Huge pits were dug at the ends of most of the tombs' entrance corridors. Archaeologists debate whether the pits were to catch rainwater and prevent it entering the burial chamber or to entrap grave robbers. They may have done both. The purpose of walls constructed on the plateaus above the entrances to some tombs is quite obvious, however; it was to divert storm water. These walls were probably quite effective initially, but over the years the stones in them were washed away, and runoff from the plateau above has been able to flow into the valley in recent years. Efforts to prevent further damage from flooding have taken the form of building—or in some cases rebuilding—diversion walls both on the plateau above and in front of tomb entrances.

Modern investigators found many tombs completely filled with debris washed in by recurrent, if infrequent, floods. The floods caused considerable damage to the decorated walls: the debris abraded the paint and the water soaked into the plaster. On the other hand, the rocky filling has prevented even worse damage in a few cases by helping to support pillars and ceilings that would otherwise have collapsed.

Much damage occurs when the floodwaters enter the lower levels of tombs that have been excavated into the Esna shale layers beneath the limestones. The clay of the shale absorbs water and expands. Tests have shown that clay minerals cause unconfined Esna shale to swell more than 50 percent when wetted; under a pressure of 2,445 kg/sq m (500 lb/sq ft) the shale still increases 12 percent in volume! This creates tremendous pressure upward on the limestone; when the shale dries out, it contracts allowing the limestone to subside. Repeated cycles of this type fracture the limestone. Pillars that have been left to support the ceilings of rooms shatter and lose contact with the ceiling; walls collapse inward. Even where the ancient excavators have not penetrated into the Esna shale, there are many natural fractures of the rock that allow rainwater to penetrate through the limestone into the shale. Furthermore, the lower layer of Serai limestone Member One also contains some layers of shale.

Swelling of the shale layer can affect limestone layers above it in another way besides upward pressure. The limestone has many vertical joints dividing the massive rock into many subsections: some joints formed by pressure from the underlying shale, some by **pressure release** as the adjacent limestone rocks were eroded away to form the wadi. When the shale gets wet, it becomes slippery. This allows the limestone sections to slide under the force of gravity downward from the parent mass of rock toward the free space of the wadi. This has the effect of opening joints even further and augmenting the process described in the last two paragraphs.

As mentioned above, three royal tombs penetrated through the base of Member One of the Serai limestone into the Esna shale below, and each tomb has suffered almost total destruction as a result. KV20, the tomb excavated for Hatshepsut descends 98 m from its entrance; about two-thirds of this distance is a corridor cut through shale to the burial chamber. This chamber is now completely filled with collapsed pillars, walls, and ceilings. KV7, the tomb of Ramesses II, should be one of the stars of the necropolis. Instead, it is almost completely destroyed and filled with flood-borne debris. The chambers of this tomb do not penetrate deeply into the shale layers but sit atop them, as if the builders had learned not to go deeper. A similar situation exists in KV17, the tomb of Ramesses' father, Seti I. Here, too, the chambers rest at the top of the shale layer. This tomb was still in excellent condition when it was opened by Belzoni in 1817, and it became a popular tourist destination. Belzoni had the deep well shaft filled in, and flooding immediately became a problem; deterioration was rapid. Recent geological studies of the 62 tombs in the Valley of the Kings have identified three more tombs at moderate risk of damage caused by swelling shales: KV5 (sons of Ramesses II), KV11 (Ramesses III), and KV16 (Ramesses I). Fortunately, for the rest of the tombs, this risk is low.

Both floodwaters and water seeping through the fractured rock walls, and even the breath of tourists, provide moisture that can mobilize the salts in the limestone. When the moisture dries, the salts recrystallize behind the paint and plaster and force them away from the wall. In many tombs, small drifts of painted plaster lie at the base of the walls. Conservation efforts to reattach the plaster—such as those undertaken in the tomb of Nefertari in the Valley of the Queens—are possible but very time-consuming and expensive.

The shortcomings of the Valley of the Kings as a place to build underground tombs arise not from the properties of limestone, per se. The limestone of the Serai Formation is generally fine-grained and hard and can make good building stone. The weaknesses come from the jointed nature of the cliffs themselves, unfortunately underlain by Esna shale, with its propensity to swell and slide. Recommendations of methods to stop the damage in the tombs and provide some mitigation have been offered by mining engineers, but mining practices cannot be applied indiscriminately, since mines are intended to operate safely for only a few decades, whereas the remediation of the tombs should ideally preserve them for many future generations.

Rock bolts, designed to retain fractured rock, cannot be used, since they will corrode in about fifty years. Reinforced concrete has been shown to undergo rapid disintegration when in contact with the salts of the Serai Formation limestones. Steel posts and beams detract from the appearance of the ancient architecture, but are probably the only way to hold up the ceiling in small chambers. It has been recommended that damaged pillars be replaced by ones carefully built of local limestone; doorjambs and some interior walls can be similarly rebuilt, especially where the surface decoration has already been lost. Such rehabilitation may not make the tomb a candidate for public viewing, but it will prevent further deterioration and allow archaeologists to continue studying it safely.

Before leaving the West Bank, we should make one last observation on the consequences of geology for human monuments. Now that the properties of Esna shale have been described, we can return briefly to Hatshepsut's temple, which was built directly on this treacherous layer. Rising ground water in the area has caused the shale beneath the temple to expand and displace some of the original temple walls as well as restored portions. Moisture is also carrying salts into the decorated blocks, with results that are all too familiar to the reader by this time. The temple restorers are going to dig out the shale foundation and replace it with one of fired brick. Credit should be given to the excellent study and restoration work carried out on this temple over many years by the Polish–Egyptian Archaeological Missions at Deir el-Bahari.

The Nile Valley from Luxor to Cairo

A s we travel north of Luxor, at first we notice little change in the Nile Valley. Limestone cliffs appear in the distance beyond the villages and fields of the floodplain. As we approach the city of Qena, the cliffs on the west bank draw closer to the river. At Qena the river bends to the west and flows between steep cliffs on either side. At Nag Hammadi, the channel resumes a more northerly course, and the floodplain grows gradually wider. Geologists have long debated the cause of this unusual bend in the river. New ideas about separate origins of the river valley north and south of Qena (see Chapter 13) may resolve the issue.

Limestone Cliffs

For nearly the entire distance from Luxor to Cairo, limestone cliffs bound the Nile Valley. The Eonile River cut this valley into a plateau of Eocene Epoch limestone during the late Miocene. The plateau was also dissected by streams running toward the Nile. During the subsequent Pliocene, the Nile valley and its tributary valleys were filled with marine and fluvial deposits. Pleistocene rivers (the Prenile and the Neonile) cut channels through this fill but left remnants of Pliocene age deposits along the valley walls and in the mouths of wadis. These rivers also deposited additional sediments on a floodplain that was higher than the one in existence today. These deposits form elevated terraces between the modern floodplain and the limestone escarpment. In some places sand and gravel of Oligocene age or later cover the high limestone plateau, but in most places it is bare and rocky.

At Luxor we noted the limestone cliffs of the Serai Formation of the Thebes Group. These were formed during the Early Eocene when a warm shallow sea—an extension of the Tethys—covered most of Egypt. By the Middle Eocene, the coastline had retreated toward the north, so that limestone formed during this interval extends only as far south as Asyut. Therefore, as we cruise north, the rock visible in the cliffs becomes younger. (The reason that we don't simply see a higher cliff with Middle Eocene rock stacked atop Early Eocene rock is that the strata dip toward the north. Thus the Early Eocene rocks do continue beneath the Middle Eocene rocks, but are submerged below ground.)

The limestones between Asyut and Cairo were formed during the Middle and Late Eocene. Geologists recognize various formations, but there is still considerable debate among them over the best set of names and subdivisions. The names and formation boundaries in Figures 9.1, 9.2, and 10.1 are taken from *The Geological Map of Egypt, 1987* for the most part, although I have retained the older usage of Mokattam Formation for most of the layers north of el-Minya. (Some geologists subdivide this layer into Gebel Hof, Observatory, Giushi, etc., but such detail is unnecessary for this book).

The rock formations differ in their date of deposition as well as in their composition and fossil content. These differences reflect variations in the conditions under which the rock formed. We can see from Figures 9.1 and 9.2 that the same formation is usually found on both sides of the Nile in a particular latitude, indicating, as we have already described in Chapter 3, that the Nile Valley was cut later than the period in which the limestone was formed.

Four important layers of Eocene limestone can be observed in the cliffs—especially those on the east side of the Nile. These are called (from the bottom and oldest up) the Minia Formation, the Samalut Formation, the Mokattam Formation, and the Maadi Formation. The limestone layers on the east bank of the modern river are thicker than the corresponding ones (formed contemporaneously) on the west side. The explanation for the differences in thickness and types of rocks of the same age is the presence of hills and basins in the underlying platform of older Cretaceous rocks. These hills and basins were produced by tectonic forces: namely, uplift and compression that caused folding. A thick layer of sediment will collect in a deep ocean basin, while a shallow area will accumulate only a thin layer. Furthermore, water depth determines the kind of deposit and hence the kind of stone formed (review Fig. 2.1). Solid limestone such as that in the Serai

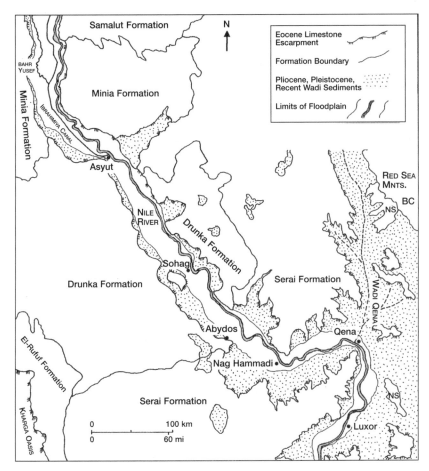

Fig. 9.1: Map of the Nile Valley from Luxor to Asyut, showing the names and boundaries of the various Formations of Eocene limestone. NS = Nubian sandstone exposed, BC = basement complex exposed.

and Mokattam Formations forms in the deep marine environment of a basin; shales and marls form in shallow water, especially where inflowing rivers bring loads of clay and silt to mix with the calcium carbonate. One of the consequences for human societies of this variation, which often occurs over fairly short distances, is that the rock in some areas is more suitable for building materials than in others.

While suitability of the rocks—based on their differing properties—was always an essential factor in determining whether to exploit a particular source, transport costs had to be considered as well. For this reason, inferior rock from a local source might be accepted rather than insisting on a superior rock that had to be moved a great distance. Probably only the kings enjoyed the wealth and power to use the best stone regardless of its transport difficulties. We find, therefore, that there were many quarries along the Nile. Those between Luxor and Asyut quarried rocks of the Thebes Group for local building projects. The Minia and Samalut Formations are both well developed in the eastern cliffs of the Nile Valley from Asyut to north of el-Minya. These limestones range from fine- to coarse-grained and contain many fossils. Both provided stone of only mediocre quality for local building, sculpture, inscribed walls, and **stelae**. Proceeding toward Cairo, another series of quarries tapped the younger limestones of the Mokattam Formation.

The Mokattam Formation is the source of stone at the famous quarries of Gebel Tura and Gebel Hof in the Cairo area. The name **Mokattam** comes from the massive cliffs of Gebel el-Muqattam of Cairo, where the formation is 130 m thick. The lowest layers of this formation contain some fossils of *Nummulites gizehensis,* but most of the rock has few if any fossils and is extremely fine grained. This made it ideal for casing stones where a smooth uniform surface was required. It could also be carved into very detailed statuary or raised reliefs. This stone was quarried extensively beginning in the Old Kingdom. The color of the stone is creamy white or gray; the color may deepen slightly as it weathers, due to oxidation of iron compounds in the rock. Evidence of quarry activities can be seen on Gebel el-Muqattam east of the Citadel, although ancient workings have been obliterated by recent activities. Stone is still being quarried from the Mokattam Formation in several locations. Some of the stone is destined for building blocks, but a great deal of it is made into lime for cement or concrete blocks or used in other kinds of industrial processes.

Limestone of the Maadi Formation varies in its depth and lateral extent. It is found in only a few places on the west side of the river. In the desert to the southeast of Cairo the formation is up to 77 m thick, but in the Giza Pyramids plateau and southward to Saqqara the formation is thinner. It has some layers of brown limestone sometimes called 'Saqqara limestone,' which was used for core blocks in pyramids or mastabas. Since it consists of a lower unit of

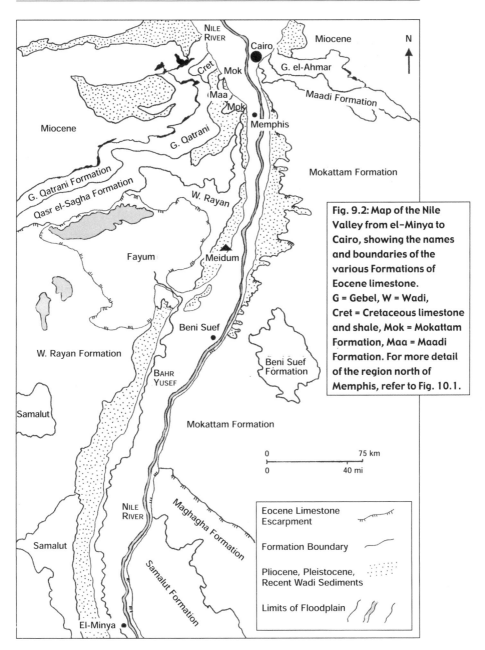

N

Fig. 9.2: Map of the Nile Valley from el-Minya to Cairo, showing the names and boundaries of the various Formations of Eocene limestone. G = Gebel, W = Wadi, Cret = Cretaceous limestone and shale, Mok = Mokattam Formation, Maa = Maadi Formation. For more detail of the region north of Memphis, refer to Fig. 10.1.

0   75 km
0   40 mi

Eocene Limestone Escarpment

Formation Boundary

Pliocene, Pleistocene, Recent Wadi Sediments

Limits of Floodplain

brown sandy limestone and an upper unit of shale and sandy shale, it was not exploited for temples or art.

North of Cairo we encounter rocks formed during the Oligocene and later. The Oligocene deposits consist of gravel, sand, and other terrestrial remains including animal bones and petrified tree trunks, which indicate they were deposited by rivers flowing from the Red Sea Mountains. At Gebel el-Ahmar, some of the quartz sand has been cemented to form quartzite. This **lithification** occurred when superheated waters, charged with silica and iron oxide, rose up along faults perhaps driven by the volcanic activity dated to that period. The same process produced the petrified (or silicified) wood found in these formations.

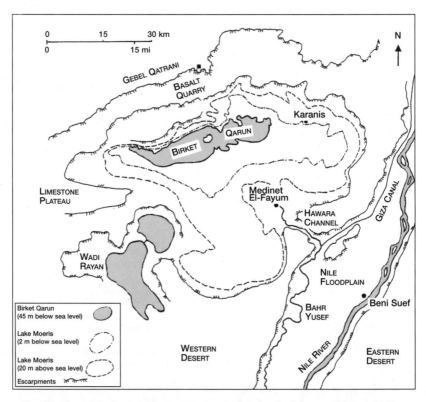

Fig. 9.3: Map of the Fayum, illustrating the shoreline of its lake at various periods.

The Fayum Depression

Viewed from the air, the lush green Fayum Depression 70 km southwest of Cairo looks like a leaf attached to the stem of the Nile Valley. The depression has an area of 1,700 sq km, and its lowest point lies at 53 m below sea level. It differs from the other depressions in the Western Desert in being connected to the Nile Valley via the Hawara Channel, through which a branch of the Nile called the Bahr Yusef flows. Irrigation water that enters the depression to be distributed across the fields eventually drains into Birket (Lake) Qarun in the northwest margin. The area and depth of the lake has varied greatly in the past—sometimes in response to natural forces and sometimes through human intervention. It currently has an area of about 200 sq km and a maximum depth of 8 m.

The processes that formed the Fayum have been much debated, but its geological history is gradually being clarified. During the Late Oligocene Epoch, the Tethys coastline lay at the latitude of Cairo. The Red Sea Mountains were being uplifted. Runoff from the northern part of this uplifted area was carried by many wandering streams that continued far to the west of the present River Nile, since there was no River Nile or Nile Valley as we know it today to receive them. These streams carried coarse sediments, some of which were deposited in a delta on the Tethys coast in the vicinity of the present Fayum. These gravels and sands were later cemented into **conglomerates** and sandstones that form the Gebel Qatrani Formation. This Oligocene formation overlies older Eocene rocks of the Qasr el-Sagha Formation, which is composed of layers of soft shale and limestone.

In the late Oligocene, tension associated with the opening of the Red Sea created fissures in the surface of the land across northern Egypt. Molten basalt erupted through these fissures and spread across the gravel beds of the Gebel Qatrani Formation. This thick sheet of dark basalt can be seen topping the steep Gebel Qatrani escarpment on the northwest rim of the Fayum.

In the Miocene, the Tethys coastline retreated even farther to the north. Instead of being submerged, the region occupied by the present Fayum and the area to the northwest of it were uplifted and formed the top of a northeast–southwest fold resulting from northwest to southeast compression at about this time. The compression may have been the first effects of the collision between the African and Eurasian Plates. Such compression and uplift would have cracked the brittle limestones.

During the late Miocene, the Mediterranean desiccation helped set the stage for the carving of the Nile Canyon—a process described in Chapter 3. When the Mediterranean refilled during the Pliocene, the Nile Canyon became a gulf of that sea, and marine deposits were formed all along the sides of the canyon at a height of 180 m above current sea level.

Pliocene deposits are found on top of the escarpments defining the eastern edge of the Fayum, but there are none within the depression itself. These particular Pliocene deposits are not marine formations but are sandy gravels that appear to have washed off the Oligocene conglomerate highlands to the west. This shows that the Fayum depression was formed after the Pliocene invasion of the Mediterranean into the Nile Valley, that is, sometime during the late Pliocene or the early Pleistocene. The date of formation can be further limited by the discovery of deposits at a height of around 44 m above sea level in both the depression and Nile Valley with flint tools thought to be at least 70,000 years old.

Debate about the forces creating the Fayum and other depressions in the Western Desert has raged back and forth between those espousing water erosion and those supporting a mainly **aeolian** (wind deflation) mechanism. Rushdi Said is among those who feel that water played an important role. He believes that rainwater seeped into joints in the limestone strata formed by tectonic forces and uplift. The acidified rainwater dissolved underground channels and caverns, a process known as **solution weathering** (Fig. 9.4). When the roofs of the caverns became too thin they collapsed, breaching the desert surface and generating rock fragments. Once the limestone layers were breached, the underlying friable shales and marls of the Fayum area were easily eroded by water and wind.

The meandering nature of the Bahr Yusef proves that it is a natural channel rather than artificial. It branches off the Ibrahimiya Canal (which replaced the Bahr's earlier direct connection to the River Nile near Asyut) and runs 330 km to the Hawara Channel. During its northward course the Bahr Yusef hugs the western side of the Nile Valley, while the main channel of the Nile flows along the eastern escarpment. The Bahr is generally 100 m wide and 4 m deep. The Bahr may represent an old river channel that was not completely abandoned after the river changed course, or it may be that in this stretch where the valley is wide, there were two major branches of the river at some date in the past. Entrance of water into the Fayum from the Bahr Yusef is regulated by sluice gates; water not diverted into the Fayum continues northward in the Giza Canal.

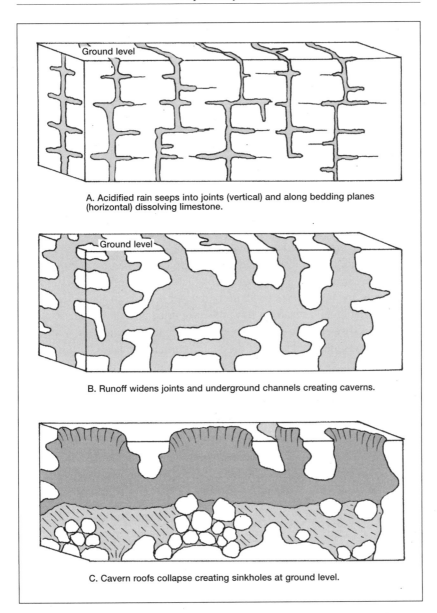

A. Acidified rain seeps into joints (vertical) and along bedding planes (horizontal) dissolving limestone.

B. Runoff widens joints and underground channels creating caverns.

C. Cavern roofs collapse creating sinkholes at ground level.

Fig. 9.4: Diagram illustrating the solution weathering of limestone that probably contributed to the formation of some of the depressions.

Borings into the sediments filling the Hawara Channel revealed that there is a deep gorge cut into the bedrock (beneath the Nile alluvium) that extends to a depth of 17 m below sea level. Rushdi Said postulates that the channel originated as a wadi draining into the Nile Valley from the elevated western desert. As the wadi cut deeper and westward, it eventually intersected the Fayum depression on the other side of the narrow plateau (height 50 m above sea level) that separated the depression from the valley, whose height outside the Fayum is 25 or 26 m above sea level. The deep channel would have been cut at a time when the sea level and hence the level of the Nile and its flood-plain were much lower than they are today.

Eventually sea levels rose, the River Nile aggraded its channel and the floodplain rose. Because of the connection between the Fayum and the Nile Valley, river water could enter the depression during the flood stage and form a lake. These high water levels created deposits of sediments on the floor of the lake within the depression and beaches around its margins. An examination of these deposits suggests that the first lake was formed with water from the **Prenile** and had a depth of 88 m (or a water level 43 m above sea level).

When the Prenile ceased to flow, this lake dried up. But the later **Neonile** brought floodwaters high enough to enter the depression and form a series of lakes. Humans inhabited some of the beaches around the lakes during the late **Paleolithic** and **Neolithic** phases. The flint tools and other artifacts recovered from the beaches have helped to date their deposits. It appears that high Nile water levels produced lakes in the Fayum several times between 9000 BCE and 3900 BCE. Around 3000 BCE a lake, known as Lake Moeris, formed and last-ed throughout the pharaonic era, but drops in its levels are recorded at around 2000 and 1200 BCE, reflecting lower Nile flood levels as already discussed in Chapter 4.

During the Old Kingdom, the lake level was probably around 20 m above sea level and nearly filled the depression. There was a free flow of floodwater from the Nile through the Hawara Channel that maintained this level. Basalt was quarried from Gebel Qatrani for the Old Kingdom pyramid complexes where it was used as pavement in many temples. Blocks of naturally-fractured basalt were moved down from the quarries along a specially-constructed road to a quay on the shore of Lake Moeris. There the blocks were loaded onto barges that could move out of the lake into the Nile Valley during the period of high flood and complete their trip to Giza via the river.

A series of low floods—already proposed for the First Intermediate

Period—allowed the Hawara Channel to fill with debris and wind-blown sand. During the Middle Kingdom high floods reappeared and Twelfth Dynasty kings decided to clear the Hawara Channel and allow the Fayum to receive some of the floodwaters. Apparently some of this water returned to the Nile Valley as the flood abated. The flow in and out of the depression was entirely automatic, depending only on the natural rise and fall of the Nile. The maximum lake level was probably close to 20 m above sea level.

Ptolemy I in the third century BCE decided that it was no longer necessary to use the Fayum for flood control. By building sluice gates in the Hawara Channel he was able to control the amount of water that entered the depression. With reduced inflow and rapid evaporation from its surface, the lake level dropped to around 2 m below sea level. The fertile land exposed on the former lake bed was put into cultivation. Fields to the south and east of the lake were irrigated by canals that diverted from the Bahr Yusef near the town of Medinet el-Fayum. After flowing across the fields, the water drained into the lake. Fields on the north shore were watered using *saqyas* to lift the water from wells, which tapped a water table maintained by seepage from the lake and canals.

The Fayum was densely populated during Greco-Roman times, when new towns were built there to house Greco-Macedonian veteran soldiers. The ruins of Karanis, one of these towns, can be visited. It sits high and dry on a ridge far from Birket Qarun, but the town lay on the lakeshore of that period. Ptolemaic innovations increased the agricultural productivity of the country and led to the highest population levels (estimated at up to 4.9 million people nationwide) at any time in Egyptian history until the end of the nineteenth century. The agricultural improvements were not made to increase the welfare of Egyptians but to provide cash crops that could be sold on the European market.

The Fayum is still one of the most productive agricultural areas in Egypt. Irrigation water enters from the Bahr Yusef, is distributed to the fields via canals, and drains to Birket Qarun. Since there is no outflow from this lake except for water lost by evaporation, it has become more saline with time. Today it has a salt concentration equal to that of seawater. The freshwater Nile fish that supported a flourishing fishing industry in the past cannot survive in these conditions, but stocking of the lake with marine mullet has protected the industry.

The productivity of the Fayum farms is limited by the amount of irrigation water that can be introduced to the depression. If water were allowed to flow in faster than wastewater can evaporate from the surface of Birket Qarun, the lake's water level would rise and flood part of the currently cultivated land. To solve this problem, a pipeline was constructed from the Fayum into the Wadi Rayan depression located southwest of the Fayum. Some irrigation water now drains into Wadi Rayan, where it has created two lakes.

Changes in the Nile's Course

The gradient of the River Nile from Aswan to Cairo today is only 1:13,000 (or a drop of only 1 meter in 13 kilometers of length). Like most rivers flowing over a floodplain composed primarily of fine silts with a small gradient, the Nile moves in a series of back-and-forth curves called **meanders**. The water flows faster on the outside of the curves and cuts into the bank; it flows slower on the inside of the curves and builds up a sandy sediment bar. Gradually the curve becomes more exaggerated until the river cuts through a loop, creating a short stretch of straight channel. The position of a given meander progresses downstream. The river may also change its channel completely, if it breaks through its levees during high water and flows in a different bed.

These motions of the river back and forth across its floodplain tend to widen that plain, since when the river is close to one side of its valley, it will be cutting directly at the valley walls. The Nile Valley was not formed by the modern river, but by a much more vigorous river under very different conditions. During the Miocene Mediterranean desiccation, the gradient of the Nile and its volume were much greater, enabling it to cut deeply into the bedrock. This deep gorge was widened not simply by the river but by other erosive forces. The modern Nile inherited this ample valley and has done little to widen it. Instead it has excavated its channel within the ancient Pliocene fill of the Miocene Canyon.

Today the Nile between the bend at Qena and Cairo lies closer to the east side of the valley than the west. In the past, its successive channels ran many kilometers to the west. Evidence of these earlier channels occurs in the form of abandoned levees of gravel that may still rise above the level of the surrounding plain. Where they have been eroded to ground level, they can be detected by boring, since the levees extend many feet into the sediments to the depths of the low water mark of the former rivers. Actually, one need not wonder as some authors have done about why the Nile has moved from west to east so consistently—just like the hems of ladies' skirts: when they are high

94

there is only one direction to go and that is down. So a channel at the extreme west side of a valley will tend to move eastward. As the volume of flow drops, the river becomes less meandering.

Since the evidence shows that for the past few hundred years and perhaps even the past few millennia, the channel has moved eastward in many places, can we project these movements backward to the early dynastic period? There is good reason to think that Memphis was on the Nile when it was founded around 3100 BCE, and it was still on the river when Herodotus visited around 450 BCE. The ruins of Memphis now lie several kilometers west of the Nile, indicating how much its course has changed.

Slightly further north, in the vicinity of Cairo, the river has undergone a steady westward shift in the past two thousand years. Cairo was not founded until 969 CE, but it lay just north of Fustat, the first Islamic capital, settled in 641 CE, and Babylon with its Roman fortress. This fortress was on the river when it was built; now it lies half a kilometer east of the river and many meters below ground level, due to silt build-up. In 716 CE the Muslims built a **Nilometer** to measure the height of the flood on Roda Island to replace the one at Memphis; perhaps the older one was rendered unusable due to changes in the course of the Nile. They also measured the rate of silt deposition at the Nilometer and found that the silt did not accumulate at a constant rate: in some centuries the build-up was rapid, as much as 65 cm, while in others it achieved only 10 cm. Rushdi Said suggests that changes in sea level may account for part of the variation. The Nile's riverbanks were stabilized at the end of the nineteenth century to prevent further changes in the river's channel within the city of Cairo.

Ancient Memphis and its Cemeteries

M ost people touring with a group on a fixed itinerary will be taken to Memphis, or as the ticket says, *Mit Rahina*. This tiny village marks the site of Egypt's capital city during the early dynasties and Old Kingdom. Memphis remained an important administrative center even after the religious capital of Egypt moved to Thebes during the Middle and New Kingdoms. The few remains that can be seen today offer no clue to the extensive walled city with the numerous temples, palaces, residences, and workshops it must have contained. Some excavation has taken place over the years, but the depth of the silt and the presence of modern villages have frustrated more extensive efforts.

The Founding of Memphis and Development of its Cemeteries

According to tradition and Herodotus (around 450 BCE), Memphis was founded by Menes, the king who united Upper and Lower Egypt around 3100 BCE. The city was located at the junction of the Nile Valley and the Delta. The ruins of Memphis lie several kilometers west of the River Nile today, which shows how far the course of the river has changed in the last five thousand years. Originally the city fronted on the west bank of the river, perhaps situated on a natural levee. Herodotus reported that Menes built a dike to protect the city from the floodwaters. We have already seen that Karnak and Dandara had massive brick walls surrounding the temples as a safeguard against flooding. The temples in Memphis were probably built of stone, but their blocks were quarried and recycled for other projects after the city was abandoned. The palaces and homes were built of mud brick throughout the dynastic period and have been lost to the floods or to *sabakh* diggers.

Cemeteries—whether they consisted of simple pit graves or elaborate tombs—were generally placed in the desert, often elevated on ridges or cliffs. There, beyond the reach of floodwaters, the burials can survive for millennia unless they fall prey to tomb robbers. Some people think that all cemeteries were on the west bank of the Nile, but towns on the east bank usually had their cemeteries in their vicinity. Thebes, a city on the east bank near present-day Luxor, seems to have been a major exception. As we have seen, the royal cemeteries along with the tombs of nobles were on the west bank. The west bank also had palaces and residential areas, however.

The Third Dynasty marks the period when both kings and high officials began to construct tombs of stone rather than mud brick. The mortuary temples and chapels contained inscriptions carved in stone and stone statues. It is the cemeteries rather than the ancient city, therefore, that provide evidence allowing us to trace the evolution from mud-brick to stone construction during its first flowering in ancient Egypt. In them, we can see how the Egyptians made use of their geological resources and coped with geological challenges. The resources consisted of rocks of various types with differing properties. The challenges included the location of useful rocks and the limitation that their properties placed on construction methods.

The Saqqara Necropolis

The earliest cemeteries of the Memphis metropolis were located at Saqqara atop a cliff of Eocene limestone (Fig. 10.1). Officials in the First and Second Dynasties built large mud-brick mastabas in a row along the western escarpment, overlooking the city in the valley below. The design of their tombs mimicked those of the kings, who were still buried at Abydos, the traditional home of the earliest dynastic rulers. The officials' graves are so grand that archaeologists at first thought they were royal tombs or cenotaphs, but the modern interpretation views them as non-royal. Royal burials probably began at Saqqara in the Second Dynasty, but these tombs have been hidden by later constructions.

The earliest tombs at Saqqara lie to the north of the Step Pyramid, built by King Djoser in the Third Dynasty. For this reason, it has been suggested that the earliest part of Memphis lay at approximately the location of the modern town of Abusir. From this location, a large wadi ascends southward to the desert plateau. Along this route are two Third Dynasty monuments—one of which belonged to Djoser's successor while the other may predate him.

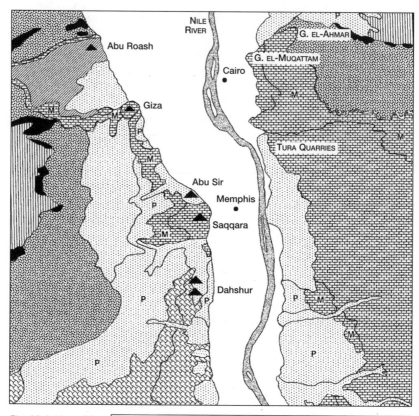

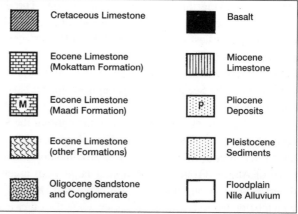

Fig. 10.1: Map of the ancient Memphis region, showing sites of pyramid complexes and quarries. Patterns indicate different rock types exposed at the surface.

Cretaceous Limestone

Basalt

Eocene Limestone (Mokattam Formation)

Miocene Limestone

Eocene Limestone (Maadi Formation)

Pliocene Deposits

Eocene Limestone (other Formations)

Pleistocene Sediments

Oligocene Sandstone and Conglomerate

Floodplain Nile Alluvium

Djoser's Step Pyramid is often described as the world's 'first stone build-ing.' I would counter that it is the earliest stone building that we know of. Excavations are already revealing earlier examples of stone architecture, showing the development of the craft. It is easy for the visitor to be so over-whelmed by the impact of the clean modern lines of the buildings that they do not notice the construction methods. The 'buildings' are nearly all dummies: most of their interiors are filled with rubble, not rooms. The rubble is retained within walls of rough stone blocks. Outside these is a casing or veneer of fine-grained, neatly finished limestone blocks.

Much of the stone for the Step Pyramid complex was quarried nearby. Limestone outcroppings are very accessible in the escarpment and in the wadis that dissect the plateau from the east and north. Although the strata are relatively thin, with dense stone separated by layers of crumbly marl or sandy limestone, they can easily be quarried into small manageable blocks. Note that rubble is a natural by-product of quarrying limestone blocks since it comes from the softer beds between the denser layers. Shaping limestone blocks also creates chips of all sizes, which is what I mean by 'rubble.' Early archaeologists referred to this material as 'rubbish,' but that term has an entirely different connotation today.

The Entrance Colonnade of the Step Pyramid Complex is based on the simple post-and-beam mode of construction. Some authors have asserted that the attached columns of the entry hall reflect the builders' unfamiliarity with stone construction and their fear that these stone columns would not support the roof. I do not agree. I think the attached columns were designed to form a series of niches into which statues were set. Everything about the architecture of this complex speaks to me of confidence in the use of stone as a building material and thus prior experience in using it. The columns were formed by stacking small drums of limestone to the desired heights; no monolithic columns were employed in this complex.

The Step Pyramid underwent several cycles of overbuilding, growing from a mastaba into a six-stepped pyramid. In the mastaba stages, the blocks were set horizontally, but in the towering steps the architect adopted a more conservative approach. The steps were constructed with a series of concentric layers, each of which leaned inward at an angle of about 78 degrees against the layer behind it. The limestone blocks averaged 30 x 30 x 15 cm in size and weighed about 35 kg—a size that two men could carry and set in place. The core blocks for the steps doubtless came from quarries very close to the site.

The casing, which is a finer-grained stone, was almost certainly brought from one of the quarries on the east bank that tapped the limestone of the Mokattam Formation. One can see Gebel Tura directly across the Nile Valley to the east.

Limestone replaced mud brick for the tombs of officials beginning in the Old Kingdom. Some of the tombs were also cut into the rock walls of quarries and wadis, although this was not as common a practice at Saqqara as it would become at Giza, where the thickness of the limestone strata facilitated such excavations.

Evidently the royal necropolis was moved from the historic site at Abydos to Saqqara some time in the Second Dynasty, in order to be close to the new capital. It is still a mystery why the kings at the end of the Third Dynasty and during all of the Fourth Dynasty abandoned Saqqara and chose successively as sites for their funerary complexes Meidum, Dahshur, Giza, Abu Roash, Giza again, and perhaps Zawyet el-Aryan. These sites cover a stretch of the western desert extending for 46 km. Various authors have proposed hypotheses to explain the different locations; these include location of king's family home (either historically or where he himself chose to build a palace), sites with previous religious significance, and location on a site that could be viewed from the religious center at Heliopolis. None of these suggestions can be completely ignored, but for none of them is there overwhelming evidence.

I think that practical considerations were more probably decisive. We have already noticed on our trip down the Nile that sandstone and limestone quarries were located at many points along the way in the cliffs overlooking the river. These quarries provided stone for local building projects such as temples or tombs for regional officials. Two exceptions to this pattern of using the nearest resource have already been noted: Aswan supplied granite and quartzite to many downstream sites, while Gebel Silsila was the source of sandstone for many of the major temples from Kom Ombo downstream and even past Luxor. The use of these quarries for royally-commissioned projects has been rationalized in the case of Aswan by showing that it was the best source of granite: the quarries were close to the river and all travel was downstream. Gebel Silsila was equally convenient to the river, and an alternative high-quality building stone was not available in the vicinity of Luxor. Admittedly, we are looking at cases that occurred during the Middle and New Kingdoms, but the logistical considerations must have been similar for the Old Kingdom builders.

In fact, the logistics were probably of even greater concern for kings who planned pyramids for their tombs. Whereas New Kingdom temples were hollow structures with courtyards and chambers enclosing much open space, pyramids were essentially solid, with only a tiny proportion of their interior occupied by passages or chambers. Thus the chief concern of the king and his architect was to find a suitable source of stone and site the pyramid as close to it as possible, while still searching for a position that would enhance its appearance. The edge of the western escarpment would seem to satisfy both of these objectives, since a pyramid at that location would be visible from the floodplain below, and the quarrymen could attack the bedrock at its exposed edge. The solid bedrock of the plateaus obviously was a better foundation for this tremendous structure than the muddy floodplain.

The more I studied this question, the more convinced I became that geology was a key and often overlooked factor in ancient planning. I noticed that the only two places where limestone of the Mokattam Formation outcrops in the face of the western escarpment are at Saqqara and Giza (Fig. 10.1.), and both were the sites of major necropolises with large numbers of limestone structures. The strata at Saqqara are much thinner than those at Giza, however, and this limited the size of blocks that could be extracted. In order to build pyramids with huge blocks, such as those used in the Fourth Dynasty pyramids, it was necessary to locate strata of limestone of sufficient thickness. Giza provided ideal rock for this purpose. I only wonder why the kings did not build there sooner.

Cemeteries at Meidum and Dahshur

The Egyptians continued to build step pyramids with inward-leaning layers of blocks for a few more generations after Djoser's Step Pyramid at Saqqara. The Meidum Pyramid was built in several phases. The first phase involved a seven-step pyramid built with inward-leaning layers; this was built over and enlarged to an eight-step pyramid. It was finally converted to a true pyramid shape by applying a layer of filling and casing stones—both set horizontally—over the earlier constructions. There has been an ongoing debate about what happened to the pyramid to bring it to its current condition. The two alternatives with the strongest support are that it has been quarried away and that it collapsed.

I believe that the Meidum Pyramid's upper steps and upper casing collapsed—some time after the Nineteenth Dynasty, since graffiti dated to that dynasty have been found within the small temple at the base that was buried by debris. Investigations in the 1880s uncovered a Twentieth Dynasty burial in the upper layers of debris around the pyramid's core, which seems to indicate the latest possible date for the collapse. I think an earthquake probably shook the structure. Since the leaning layers of stone were not bonded to one another, it was possible for the outmost layers to simply slide off the inclined inner layers. In its current ruinous state, this pyramid has provided more information about methods of pyramid construction than other more intact ones.

The first pyramid built as a true pyramid from start to finish was King Sneferu's Bent Pyramid at Dahshur, closely followed by his Red (or Northern) Pyramid, also at Dahshur. If you are interested, there are many books on pyramids that discuss possible reasons why the angle of the Bent was changed. I am going to focus on the stones of the two monuments. We do not know why Sneferu, first king of the Fourth Dynasty, chose Dahshur for his funerary complex, but I can suggest a reason for locating the pyramids just where they are once the general locale had been selected. At Dahshur, the best Eocene limestone was accessible to the west of the pyramid sites. If the monuments had been placed closer to the cultivation, as the Twelfth Dynasty mud-brick pyramids were, the stone would have had to be transported farther from the quarries to the building site. The Bent and Red Pyramids are mammoth structures with volumes equivalent to 55 percent and 65 percent respectively of the Great Pyramid, so a local source of core blocks was essential. Quarries west of Dahshur would have used limestone of the Wadi Rayan or Qasr el-Sagha Formations, which are both much coarser than Mokattam rocks.

The Red Pyramid has lost most of its casing blocks, except for a small section on the lower east face. A chemical deposit on the surface of the exposed core blocks causes the distinctive reddish color. Many limestones contain iron and/or manganese oxides at low concentration. These compounds can be dissolved by atmospheric moisture. As the moisture evaporates it carries the oxides to the surface and leaves them on the surface of the rocks. Streaks of color derived from oxidation in the core blocks can be seen running from cracks in the casing blocks on the Bent Pyramid. Except for these streaks, the casing blocks on the Red and on the upper part of the Bent are pale tan, not reddish. Evidently both the Bent and the Red were cased with limestone lacking these impurities, or at least the casing contained less of them. Blocks that have been recently removed from under mounds of rubble are cream colored—showing that the discoloration results from exposure to the atmosphere. The fine limestone casing blocks for both pyramids likely came from quarries on the east bank. The casing represents only about 5 percent of the volume of a pyramid, so this relatively small quantity of higher quality stone could be obtained at a distance if necessary.

The Giza Plateau

The word *Giza*, in most travelers' minds, is synonymous with pyramids. This is easy to understand, since the Great Pyramid is located there. It is unfortunate, however, that most travelers do not have or take time to visit the vast array of smaller pyramids, tombs, and temples that crowd this huge necropolis. It has much to teach us about ancient Egyptian art, architecture, religion, bureaucracy, and construction methods. It also provides material for a number of lessons on the impact of geology on society. At Giza, the use of limestone for monumental architecture reached its peak.

The Giza Plateau contains the same sequence of Eocene limestones as found in the cliffs east of the Nile, but there are important differences between them. We explained in the last chapter how rocks formed contemporaneously but at different locations could differ in thickness and composition because of regional variation, that is hills and basins, in the underlying foundation on which deposition was occurring. Different sediment sources, as well as different water depths, contribute to variations in composition. In general the limestone strata at Giza are thinner than those in Gebel Tura or the Muqattam Hills; they also tend to be coarser and contain more fossils. This means the deposition basin was originally deeper where the Muqattam Hills now are. The fact that the Muqattam Hills are now at a considerably higher elevation than the Giza Plateau is the result of uplift in the post-Eocene period. The Giza Plateau was also subjected to a gentle folding, so that the strata there now dip slightly (about 3 degrees) to the southeast.

When sedimentary rocks form, they trap samples of the marine life present at that time. Since organisms evolve over time, geologists use these changes to help them date the rocks as well as to identify rocks from separate locations that were formed at the same time. Many of the limestones formed during the Eocene contain the remains of tiny marine organisms called nummulites. These creatures were composed of a single cell with a protective shell (called a test) made of calcium carbonate. The shape of this shell was a flattened disk, and in some species it grew to several centimeters in diameter. One of these species is so abundant and so characteristic of Egyptian limestone in the Cairo/Giza area that it was named *Nummulites gizehensis.* We can see samples of this limestone underfoot as we walk around the Great Pyramid. This coarse local limestone was quarried and used as core blocks in all the Giza pyramids.

The main quarry that provided the core blocks for the Great Pyramid is located to the southeast of this pyramid and to the east of Khafre's Pyramid (Fig. 10.2). It is thought that a ramp ran from the quarry up to the Pyramid and around its four sides during the construction process. Again, this is a topic that has been discussed in many books and can be pursued if you are interested. Other quarries were opened as needed to the southeast of the other pyramids and mastaba fields. In each case the quarrying probably began at the southeast, at the edge of the escarpment and worked toward the north and west. The location of the ancient quarries can be identified by the rock faces still visible along the north and west sides of the worked areas. Rock-cut tombs were cut into these cliffs in the Fifth and Sixth Dynasties.

A good place to see evidence of the method of limestone quarrying is on the pavement northwest of Khafre's Pyramid. At this location, rock had to be removed to level the site before pyramid construction could begin. Economically, this rock was removed in blocks that were then used to build up the southeastern corner and core of the second pyramid. We can see how the quarrymen used flint picks to cut channels wide enough to stand in to form the sides of the separate blocks. The blocks removed here were probably at least a meter tall, so the trenches between the blocks were cut that deep and then about 20 cm deeper.

To remove a block from its pedestal of rock, a deep groove was cut beneath the block on one or more sides. Then the workers placed wooden levers in the grooves and pried upward. The block would break free of its pedestal along the natural **bedding planes**. Such bedding planes are characteristic of sedimentary rock and mark places where changes in conditions during the sedimentation process resulted in a change of rock composition. The workers tried to utilize the natural rock strata as efficiently as possible: extracting blocks as large as possible from the compact strata and using the softer layers as separating planes.

As noted, the strata of Eocene rock composing the Giza Plateau are nearly flat, with only a slight dip. It is much easier to quarry sedimentary rocks when the bedding planes are horizontal than if they have a large tilt. Also, the limestone forms the bedrock of the plateau, so no overburden had to be removed. Another geological feature that can be exploited while quarrying is

104

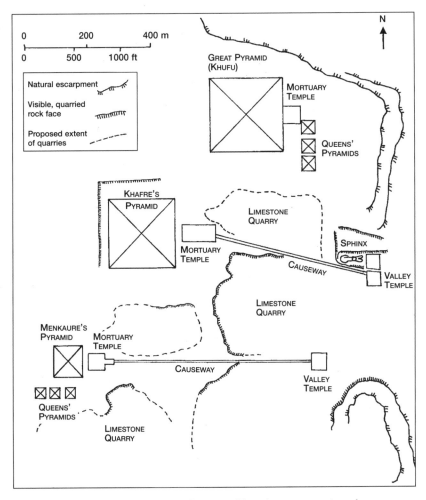

Fig. 10.2: Map of the Giza Plateau with major monuments and the locations of limestone quarries.

vertical joints that separate solid masses of rocks into smaller units, and the Giza limestone is well jointed, but not shattered. In short: Giza was an ideal quarry site and was efficiently worked by the ancient Egyptians.

Limestone was used for all the casing blocks on the Great Pyramid, and those above the first course on Khafre's Pyramid, as well on the top half of Menkaure's Pyramid. For the sake of appearance, as well as to achieve tight joints, these blocks had to be of a finer grained limestone. No such fine-grained limestone was available at Giza, so the casing blocks had to come from the Tura quarry on the east side of the river. There the Mokattam Formation has such thick strata that blocks of almost any size desired can be quarried.

The blocks were transported to the Giza Plateau on boats that probably moved along canals leading 10 km from the river to the escarpment. Egyptologists differ on whether the Nile flood would have provided enough draft to allow boats to sail overland during the months of inundation. Herodotus, claiming to be an eyewitness (around 450 BCE), reported: "When the Nile overflows, the whole country is converted into a sea, and the towns, which alone remain above water, look like the islands in the Aegean. At these times water transport is used all over the country, instead of merely along the course of the river, and anyone going from Naucratis [a Delta town] to Memphis would pass right by the pyramids instead of following the usual course...." Karl Butzer, who is an expert on the hydrology of the Nile, doubts that the typical flood of around 1.5 m would have been deep enough for heavily loaded stone barges. It has also been proposed that the Nile or a branch of it ran close to the foot of the Giza escarpment during the Old Kingdom. Further study will be required to test this hypothesis.

Where a few casing blocks remain on the northern base of the Great Pyramid, we can see that they were made of fine-grained Tura limestone set with very tight joints between them. These joints contained gypsum mortar and were probably essentially watertight. The infrequent rainfall would have run off the surface of the pyramid. Since the casing blocks have been removed, the coarser core blocks have been subject to weathering. The core blocks come from strata of limestone with different characteristics that have affected their resistance to weathering processes. Some of the blocks were cut from a hard dense limestone that still retains marks of the quarrying tools. The superiority of this rock was evidently recognized, since it was used especially as corner blocks of the core on each course. These blocks have weathered very little compared to other core blocks, since the casing was stripped off.

Other core blocks were taken from strata of soft shaley or marly limestone. These have weathered in an interesting manner. Rainfall or even nightly condensation dissolved some of the calcium carbonate in the limestone. During the day, moisture was drawn upward to the top surface; when it evaporated the calcium carbonate was redeposited. This had the effect of strengthening the top surface at the expense of the inner portion of the block from which the calcium carbonate was withdrawn. Wind erosion had little effect on the top surface of the block, but the softer portions were worn away, creating deep niches. When the pyramid could still be climbed by intrepid tourists, such eroded blocks could give way underfoot. Khafre's Pyramid appears to have been constructed with core blocks of a very erosion-prone limestone. Tiny flakes and chips that have spalled off the blocks cover its surface, nearly obscuring the individual courses. Near the top, just beneath the remaining casing blocks, the core blocks appear crisp and clean. Evidently this section of casing was removed in the more recent past and little weathering has occurred since that time.

It has been mentioned before and will be emphasized again in later chapters that the climate of Egypt was very wet at several periods since the Eocene, when the limestone was formed. In a humid climate with ample precipitation, limestone bedrock undergoes **solution weathering** (see Fig. 9.4). The Giza Plateau displays many features consistent with this mode of weathering, including wide fissures and underground cavities of various dimensions. Some of these have been identified by recent technical studies; but the ancient builders encountered one as well. This is the so-called grotto in the shaft cut under the Great Pyramid that may have been intended to serve as an escape route for workers who blocked the pyramid passages after the funeral. Vertical fissures and some solution channels can be seen in the bedrock wall northwest of Khafre's pyramid. Another huge fissure that cut across the back of the Sphinx was repaired during several episodes of restoration. Since the Giza area has been hyperarid, except for occasional cloudbursts, since around 2500 BCE, other forms of weathering have been more relevant to the integrity of the monuments there.

While the vast bulk of the stone used at Giza was limestone from the immediate vicinity or across the river, other rock types were also employed. If you have visited the granite quarries at Aswan, it is interesting to observe how much Aswan granite was used at Giza in various structures. In the Great Pyramid, the King's Chamber is completely lined with granite. The floor blocks and walls of this chamber were built within a strong shell of limestone blocks. The ceiling is formed of horizontal granite beams 8 m long, each weighing 50 to 70 tons. The Great Pyramid offers three other interesting examples of the use of granite. It was used for the three portcullises and the walls of the chamber housing them. The walls were made of granite rather than limestone to prevent tomb robbers from simply chipping away the wall holding the portcullis in place. The plug blocks at the bottom of the Ascending Passage are granite for the same reason, but intruders circumvented them by cutting through the limestone walls. Khufu's sarcophagus was dark gray granodiorite. Damaged by early tomb robbers, it also suffered from nineteenth-century tourists, who liked to chip off pieces as souvenirs.

In the Second Pyramid, Khafre used granite for his sarcophagus, to line the Descending Passage of his pyramid as far as the granite portcullis (more protection against tomb robbers), and for one course of casing blocks. The rest of his pyramid is limestone. He seemed to have preferred to lavish granite in areas that were more visible. Both his Mortuary Temple and his Valley Temple were built with gigantic limestone core blocks and cased with granite. The Mortuary Temple at the base of the pyramid's east face is mostly ruinous and seldom visited, but a trip to the Valley Temple is very worthwhile. The interior of this temple has been restored; it was badly damaged when the architraves were pulled down to extract from them the large copper cramps that had been placed in them. Each cramp was butterfly shaped and rested in grooves cut into two adjacent architraves. Since they each contained about 25 kg of copper, they were much sought after.

Menkaure's Pyramid, although the smallest of the three at Giza, made the most extravagant use of Aswan granite. At least 16 courses of the casing employed this stone, while the rest were built of Tura limestone. Inside, the Descending Passage and portcullis chambers and portcullises were built of granite. The burial chamber has a granite floor, sidewalls, and a gable roof composed of pairs of granite beams. Menkaure's temples were to have been cased in granite, but only a few blocks had been set when the king died. His successor finished the job hastily in mud brick, plastered and painted to hide the economy. Many granite casing blocks have been tumbled off the pyramid by stone robbers. The line of chiseled holes in many of the blocks reveal that the vandals date to the Roman Period or later. More ancient pillaging also occurred. An unfinished double statue, possibly of Ramesses II, was discovered in the 1990s under a pile of rubble. The granite block had been taken from one of the smaller Queens' Pyramids. The statue was abandoned when it developed a crack.

Granite was not the only exotic, non-local, stone employed at Giza. A visit to the remains of Khufu's Mortuary Temple on the east face of the Great Pyramid provides an opportunity to see the basalt pavement of the courtyard.

Careful examination of these blocks has revealed evidence of saw cuts, made by large copper saws and quartz sand. It has been suggested that basalt was used in this application because the black color symbolized the earth, or the god Geb, who had a role in the burial ritual. I have wondered how long it took the priests to discover that black rock under a desert sun can reach temperatures of more than 60° C (140° F).

By contrast, Khafre's Valley Temple has a floor of travertine (improperly referred to as 'alabaster'). This is a very attractive, but soft, rock that would normally not be a good choice for a floor. Again the choice may have been made for reasons of symbolism rather than practicality. The undecorated granite columns and beams of the courtyard were built in the simple post-and-beam style. The temple was roofed with horizontal beams resting on these architraves. Slits beneath these roof beams admitted sunlight into niches containing statues of Khafre executed in a beautiful, hard, bluish gabbro and anorthosite gneiss from the quarry northwest of Abu Simbel. One complete statue and fragments of others were found dumped (hidden) in a shaft in this temple and can now be admired at the Egyptian Museum.

Besides offering a veritable 'geology museum,' Giza also provides examples of the problems facing the ancient builders and how they solved those problems. One of the challenges to pyramid architects was how to roof the interior passageways and chambers in a way that would withstand the pressures of the vast bulk above. Three different roofing systems were used at Giza: horizontal roofs, corbelled roofs, and gabled roofs. The horizontal ceiling employs the ancient post-and-beam method of construction. Corbelling too was an old method first employed in mud brick construction and continued in stone buildings. But the gable roof seems to have been used in monumental architecture for the first time in the Great Pyramid.

The architect could not simply design a ceiling based on esthetic qualities but had to take into account the properties of the building material to be employed. In general, stone has enormous compressive strength (resistance to crushing), but it has less tensional strength (resistance to stretching) than a material such as steel. Horizontal ceiling beams have to support their own weight across the space below as well as carry the weight of anything lying above the ceiling (Fig. 10.3). Such blocks experience great tensional forces along their lower sides that may easily exceed the tensional strength of the stone employed. There is a distinct limit on the width of the chamber that can be roofed with horizontal stone beams. That width is reduced even more if a large superstructure must be supported. Granite beams can span a wider space than limestone ones of the same cross-sectional dimensions, because granite has greater tensional strength than most limestones.

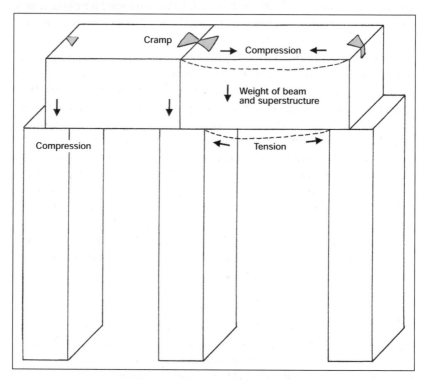

Fig. 10.3: Diagram of the post–and–beam construction technique, showing the forces involved.

In the Great Pyramid, the King's Chamber has a horizontal roof of massive granite beams—each more than 8 m long. This great length may have been the limit of what the ancient architect thought was prudent. It is unnerving for a visitor to observe that each of these beams has been cracked near the south wall, probably by an earthquake or settling of the load above. Above the visible ceiling are four more small chambers, called weight-relieving chambers, although engineers question their effectiveness in such a role. They too are formed of granite blocks for sidewalls and roofed with granite beams. Above these relieving chambers built of granite is another low chamber roofed by pairs of limestone beams set as a gable. This gable is the real weight-relieving component of this system.

Corbelling was the means used to form the upper walls and part of ceiling of the Grand Gallery. This was the method used earlier to roof all the chambers in the Meidum Pyramid and in the Bent and Red Pyramids at Dahshur. It was thus a well-understood technique by Khufu's reign. In building a corbelled ceiling, one begins by setting the first ceiling block so that it overhangs the wall blocks by only a few centimeters (Fig. 10.4). The next blocks overhang a bit farther, and so on, until the space above the chamber is gradually closed from both sides. The chief advantage of the corbelled over a horizontal ceiling is that the stones forming it are in compression and experience no tension. A corbelled ceiling could in theory be used to span a room of any width, but the wider the room, the higher the ceiling becomes. This may explain why the ceiling of the two-meter wide Grand Gallery is actually closed for the last one meter by horizontal ceiling beams.

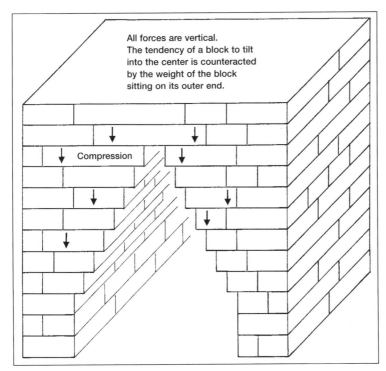

All forces are vertical. The tendency of a block to tilt into the center is counteracted by the weight of the block sitting on its outer end.

Compression

Fig. 10.4: Diagram of corbelled ceiling, showing the forces involved.

Two examples of gable ceilings occur in the Great Pyramid: in the topmost relieving chamber above the King's Chamber and in the so-called Queen's Chamber. Two superimposed gables were placed above the original entrance on the north face. Removal of casing blocks and some core blocks makes these last gables visible. The gable has all the strength of a true arch, even though it consists of only two components (see Fig. 10.5). Evidently the ancient builders felt confident of this new development, since the gable was used as the primary method of constructing ceilings over chambers in all pyramids built after the Great Pyramid. Modern explorers found that the roofing beams in the Queen's Chamber are set with less than half their length exposed in the chamber, while the rest of the length extends into the wall beyond the chamber. Each beam has a rectangular cross section and is set on its narrow side. This is the strongest way of setting a beam. Whether the ancient builders did this intuitively, or learned through trial and error, we will probably never know.

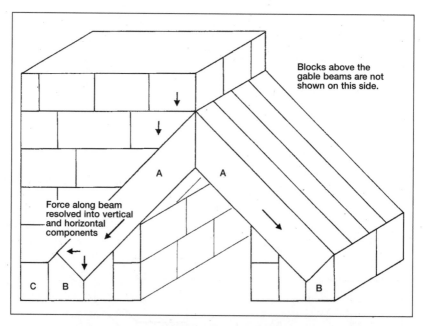

Fig. 10.5: Diagram of a gabled ceiling, showing the forces involved.
A = gable beams, B and C = core blocks.

The weight of the superstructure above the gable creates a force along the sloping gable block (A in Fig. 10.5). If this force is resolved into its vertical and horizontal components, we see that the vertical force compresses the blocks. The outward horizontal force is countered by the block B on which the gable end rests and the blocks (like C) set next to it. In Khafre's Pyramid and some others, 'blocks' B and C are cut from the bedrock and hence are immovable. In other pyramids, the gables rest on masonry blocks of the pyramid's core, and only the combined weight of this surrounding masonry contains the outward force. The gables of some Sixth Dynasty pyramids have slipped because the buttressing core blocks shifted.

The Age of the Sphinx Controversy

One of the themes of this book is that archaeologists and geologists can form a productive partnership in studying ancient sites. The Sphinx controversy seems to present an occasion in which the two disciplines came into conflict. In attempting to answer the question of when the Sphinx was carved, most Egyptologists answer: "During the Fourth Dynasty, or about 4,600 years ago." But an American geologist and other authors have suggested dates ranging from five to ten thousand years before the present. You may have read articles or seen television programs devoted to this issue. Unfortunately, some of these articles and programs give great prominence to the supposed evidence for an 'old Sphinx' while ignoring the weight of evidence against such a date.

The Sphinx is carved from the natural Eocene limestone bedrock of the Giza escarpment. It sits at the east side of the Giza Plateau, almost at the level of the valley and not on the top of the Plateau itself. The elevations of the floodplain, Sphinx, and Great Pyramid on the Plateau are respectively 17 m, 20 m, and 70 m above sea level. The Sphinx sits in a rectangular depression, called the Sphinx Enclosure or Ditch, from which the bedrock has been removed to form the nearly vertical sides of the enclosure on one side and the leonine body on the other. The gently sloping limestone strata can be traced from one side of the enclosure across the Sphinx into the other sidewall of the enclosure. Only the head protrudes above the level of the surrounding bedrock.

The Sphinx was carved into three layers or members of the Mokattam Formation of limestone. Member 1 at the base is quite dense, while Member 2, in which the body is carved, has alternating layers of dense limestone and softer marls. Member 3, at the top, has two sublayers: one soft and one very dense. The neck was carved in the softer layer and has suffered extensive erosion; the head was carved in the densest layer—as a result the head of the Sphinx has experienced less weathering than any other part. Whole chunks of material, isolated by joints in the rock, have fallen off the neck and shoulder. The body is also badly eroded with the softer layers receding deeply. Limestone blocks have been added over parts of the body during several different episodes of restoration. The neck and back of the head have been reinforced with modern concrete.

Egyptologists date the carving of the Sphinx to the Fourth Dynasty, most likely during Khafre's reign, on the grounds that the Sphinx is integrated into the plan of Khafre's funerary complex. The north side of his Valley Temple and his causeway, which runs at a slight angle along a ridge of rock left between quarries, exactly parallel the edge of the Sphinx Enclosure. Stones were quarried from the Sphinx Enclosure and used to form the core of the Sphinx Temple just to the east of the statue. The construction methods of this temple are identical to those used in Khafre's Mortuary and Valley Temples: the walls were formed of massive limestone core blocks and faced with granite casing blocks. This is different from the method used in the Khufu Mortuary Temple, in which thick walls were formed with two outer layers of fine limestone with a rubble filling. The Sphinx Temple and the Sphinx are believed to have functioned to honor the sun god, Re. Recent clearing of the Sphinx Enclosure down to bedrock revealed that quarrying had not completely removed the rock in the northwest corner. On top of the remaining rock were quarry marks identical to those seen to the northwest of Khafre's pyramid, where trenches were cut to isolate blocks. Fourth Dynasty pottery was found in the debris removed during this clearance.

The Sphinx's age thus seemed firmly settled until geologist Robert Schoch offered another opinion based on his study of the erosion patterns on the statue and results of some other tests. His conclusions were picked up and given wide dissemination by a group of people who use the older Sphinx as a centerpiece of their theories alleging that an advanced civilization had occupied Egypt as long ago as 7000 to 5000 BCE. They have also asserted that the core blocks of the Sphinx and Valley Temples—which clearly were quarried in the Sphinx Enclosure—belonged to this earlier society. In addition to ignoring archaeological evidence at Giza implicating Khafre, these people also ignore archaeological traces of actual Predynastic societies in Egypt. None of these traces provides any evidence of technology or ideology that would have led these societies to carve the Sphinx.

Schoch placed great emphasis on the deep and rounded weathering of the Sphinx's body and the south and west walls of the Enclosure. He interpreted the rounded profile and the deep fissures as "precipitation-induced weathering," i.e., the result of rain falling directly on these areas or from runoff. He tried to contrast the severe weathering on the Sphinx with the lesser weathering on certain Fourth Dynasty tombs to prove that the Sphinx was water weathered while the tomb facades were eroded by the wind. But, in fact, the tombs he cited for comparison were carved in a higher and denser layer of limestone, with entirely different properties than that of the Sphinx. Other tombs located in the same layer of stone as the Sphinx do display similar deep and rounded erosion. Schoch's attribution of various wide fissures to running water in a post-Sphinx era rather than to ancient pre-Sphinx pluvials is not persuasive. To permit the Sphinx to be drenched by rainwater, Schoch wants to move its creation back into the Holocene Wet Phase that lasted from about 10,000 to 5000 BCE (see Chapter 4).

Since the core blocks of the Valley and Sphinx Temples were obviously quarried from the Sphinx Enclosure, Schoch pushes their date of construction back to match an older Sphinx. He credits Khafre only with putting a granite casing on the old core blocks that have undergone erosion. In doing this he misinterprets the ancient method of placing a granite casing on limestone. The limestone was cut to fit the irregular backs of the granite casing blocks, not the other way around. Thus the erosion on the core blocks could postdate the stripping of the granite casing but not precede its placement.

To counter Schoch's geological arguments, other geologists have made the following points. The major fissure that runs across the back of the Sphinx can be traced to continuations in the enclosure walls in a manner that could not be explained if each of these areas had weathered independently. It is almost impossible to estimate the past rates of weathering, but even under the present arid conditions, intensive weathering has been recorded. Cloudbursts occur in the Cairo area every few years, and even today pieces of stone spall off the surface of the Sphinx and the Enclosure and form piles of debris. In fact, Karl Butzer has recently found evidence in the area southeast of the pyramids, where Mark Lehner is excavating a Fourth Dynasty industrial complex, indicating that Giza had frequent drenching rainfalls during the Old Kingdom and immediately thereafter.

The weathering processes most affecting the Sphinx include the solution of calcium carbonate by rainwater or runoff, the expansion of clay minerals, which are found in the softer layers, and subflorescence—the repeated formation of salt crystals that crack off tiny flakes of stone. The daily variation of temperatures contributes to the latter activity. At night, moisture condenses on the surface of the limestone (an event promoted by the hygroscopic property of salts, which means they absorb water); some of this moisture is drawn into the pores of the stones by capillary action (an event promoted by the geometry of the pore spaces): this provides moisture to dissolve the salts in the interior of the stone. In the morning, sunlight warms the rock and the moisture starts to evaporate. Salt-saturated water seeps back to the surface; when it evaporates, the salts recrystallize and force the rock grains apart. This same process has been referred to in earlier chapters; it is described here in greater detail, since it turns out that certain layers of the Sphinx and surrounding Enclosure walls are particularly susceptible to this process by virtue of their mineral content and microscopic geometry. Large flakes created in this fashion fall off the Sphinx of their own weight. Finer particles are removed by wind and especially wind-blown sand.

The erosion processes described above work best when the Sphinx is exposed and not buried in the sand. Yet for much of its history only the head of the Sphinx was exposed. The Enclosure is closed on the east by the Sphinx Temple and so acts as trap for wind-blown sand. Some weathering would continue under the sand layer, since the sand and its contents were repeatedly soaked by the occasional cloudburst and then dried out again. As the marly layers of the limestone were wetted with water seeping from the wet sand, the clays expanded. As the sand dried out, it drew water from the marly layers, allowing the clay to shrink; and also drew salt-saturated water to the surface where the salts crystallized and forced loose tiny pieces of rock. Removal of the weathered material was most effective, however, when wind could flow over the statue.

The Sphinx Enclosure was cleared of sand and repairs were made to the Sphinx several times since the Old Kingdom. A stela found in front of the Sphinx records one such clearance in the reign of Tuthmosis IV during the Eighteenth Dynasty. This granite stela itself shows the same kind of weathering processes described above. Ground water seeping upward into the stela and then evaporating has left salt crystals that caused the inscribed surface of the stela to flake off. The damage is greatest closer to the ground. Masonry restorations to the Sphinx seem to indicate additional clearances during the Twenty-sixth Dynasty and the Roman Period. These restorations involved placing limestone blocks over the legs of the Sphinx; its sides and back were not covered. After each restoration, the Enclosure gradually refilled with sand. Since wind-blown sand only moves a meter or so above ground level, the area of the Sphinx scoured by the sand would shift higher and higher as the sand level in the Enclosure rose. The neck and head are above the level of the Enclosure and so remained above the sand. The neck was perfectly positioned, however, for the maximum weathering and erosion.

I believe that the Sphinx controversy provides an excellent example of the workings of science and the contributions that geology can make to archaeology. By examining different kinds of evidence the two disciplines can be mutually correcting. In the case of the Sphinx controversy, the apparent clash stimulated geologists to examine weathering processes more carefully and look for additional evidence.

The Nile Delta

After flowing northward from Aswan for more than 800 km, the River Nile escapes the rocky confines of its narrow valley. About 20 km north of downtown Cairo, the river splits into two separate branches, called **distributaries**. In the past there were more than two branches, which spread the river's sediments into a broad fan-shaped region that Herodotus called the 'Delta' because of its resemblance in shape to that Greek letter. The roughly triangular region is 270 km wide at its Mediterranean shoreline and 160 km from north to south. The structure and history of the Delta is closely linked to the history of the River Nile.

Formation of the Delta

For the past several hundred million years, the northern section of Egypt has been a coastline where the African continent bordered the Tethys (later the Mediterranean) Sea. This region has nearly always been submerged, as the sea repeatedly transgressed farther over the country and retreated. As a result, from the early Mesozoic to the mid-Miocene, a thick layer of sedimentary rock formed almost continuously and covered the northern part of Egypt. During the late Miocene the land was uplifted, and these rocks formed great cliffs along the Mediterranean coast. When the Mediterranean dried up about six million years ago, a vigorous river (that Rushdi Said calls the **Eonile**) began to cut a valley more than four kilometers deep through these cliffs. Sediments from the coastal cliffs as well as from farther upstream were deposited offshore in the gradually waning ocean basin.

In the Pliocene, the Mediterranean Sea refilled. The offshore sediments were submerged by seawater and covered with marine deposits. Seawater also flooded the Nile Canyon, allowing marine deposits to form in the

118

canyon. When the canyon was about one-third filled with marine deposits, another river, the **Paleonile**, began to flow from the south, carrying sediments to the delta region. Although this river flowed for about two million years, it carried only fine-grained sediments that were mostly deposited along the sides of the delta. Then it appears that no water flowed in the Nile for a long period.

Finally, about 800,000 years ago, a mighty river, the **Prenile**, began carrying huge loads of gravel and sand northward. These sediments were laid down on the surface of the expanding delta region to immense depths. The area of the delta grew to about three times its current size and rose higher than the modern Delta. Rushdi Said thinks the Prenile River flowed for several hundred thousand years, and then it nearly disappeared as the climate in its drainage basin became arid. Eventually, when rainfall increased again, the modern River Nile, with its smaller flow and load of suspended silt, appeared. A linkage to the equatorial African lake district and Ethiopia may have increased the size of the drainage basin, resulting in a greater flow of water.

In the period since the huge Prenile Delta was formed, the sea level has fallen several times to more than 100 m below its present level. This has corresponded to the formation of glaciers in the northern hemisphere. At these times of low sea level, the Nile distributaries eroded deep channels into the surface of the Delta. Between the channels, ridges of sand remained. When sea levels rose, the river began to fill the channels and spread silt across the surface. Today, those silt layers range from 10 to 40 meters in depth. In spite of this accumulation of silt, the sand ridges have not been completely buried. In some places they protrude above the modern surface, and are called **turtlebacks**. In historic times, these ridges provided sites to build villages that were safe from the annual Nile floodwaters.

Sand dunes and shallow lagoons characterize the Delta coastline east of Abu Qir. The sand is derived from sediments carried by the Nile. These sediments are redistributed by the ocean currents moving from west to east; they are also blown back on shore by the prevailing north/northwesterly winds. Where the Nile distributaries emerged into the Mediterranean, they dropped sediments that elongated their channels. Between these narrow delta extensions, the coastline was recessed into curved beaches. The offshore currents moved the sediment into long sand bars parallel to the coast. These bars eventually enclosed lagoons between themselves and the curved coastline beyond. At first such lagoons were connected to the sea and contained salt water. But as the northwest winds blew more sand onshore and raised the dunes, the connection with the sea was severed in some cases.

The Delta in Historic Times

The Delta probably had approximately its current dimensions throughout pharaonic history, but until the Greco-Roman period the low area along the coast consisted of dunes, lagoons, and marshes, making it unfit for permanent habitation. Farther south, on higher ground, the land was suitable for grazing, cultivation, and the building of towns. Natural basins that were flooded during the inundations could be used for grazing or annual crops at other times of the year.

The courses of the Nile distributaries changed from time to time—with even greater frequency than the course of the main River Nile farther upstream. These course changes altered the number of branches and the location of the mouths. Attempts to reconstruct the Delta topography during the Predynastic period have depended primarily on geological studies of ancient channel deposits; these indicate that there were three major branches of the Nile, comparable to the later Rosetta, Sebennitic, and Damietta branches. Settlement remains help reveal the locations of former Nile distributaries during later periods, since most villages and cities were built on a branch of the Nile.

Karl Butzer describes three periods of colonization in the Delta during pharaonic times. In the Old Kingdom, many new estates were established in the southern portion. Some of these estates belonged to the king, while others belonged to high officials. The estates provided goods for the living and for the maintenance of funerary cults. Farming activities included cattle raising, market gardens, orchards, and especially vineyards.

The Hyksos invasion of the Second Intermediate Period focused the Egyptians' attention on the Levant—a focus that increased settlements in the eastern Delta during the New Kingdom. Egyptologist Manfred Bietak has excavated at Tell el-Daba, finding ruins of the Hyksos capital of Avaris, which was located on the Pelusiac branch of the Nile. This city continued to be used by the Eighteenth Dynasty after the Hyksos had been defeated. But in the Nineteenth Dynasty, Ramesses II (reigned 1279–1212 BCE) had the city of Pi-Ramesses built north of Avaris (near modern Qantir). This capital was also located on the Pelusiac branch, which provided easy access to the

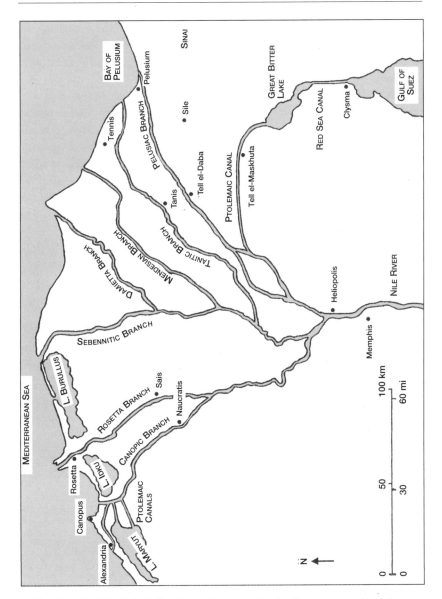

Fig. 11.1: The Nile Delta in the first century BCE, showing the names and locations of known Nile distributaries. Some historic cities are indicated, although not all of them were in existence in this particular period. L = Lake.

Mediterranean and the Near East. The Pelusiac branch had a tendency to change its course, and declining flows in this branch forced Ramesses' successors to abandon the city around 1200 BCE. It was succeeded as the capital by Tanis, on the Tanitic branch.

The Pelusiac branch managed to reassert itself at a later period, however. In fact, as the easternmost branch of the Nile, opening to the Mediterranean east of modern Port Said, it offered a route between Egypt, Sinai, and the Levant. In later Egyptian history it was the path followed by conquering Persians, Greeks, and Arabs, since boats carrying provisions could float while men marched on the desert border. The city of Pelusium at the mouth of the Pelusiac branch was a thriving city during the Greco-Roman period. Archaeologists have only recently begun excavations that reveal its size and importance as a port and trading center. It also had a number of industries producing goods for export; these included textiles, fish sauce, and salt. The population was supplied with the usual Classical urban amenities such as temples, theaters, baths, and racetracks.

During Ptolemaic times, a period of sea level recession, several coastal cities, including Alexandria, were founded or expanded. These settlements reflected important geological features (location of distributaries and contemporary sea level) as well as the orientation of the Egyptians toward the outside world. We will examine these factors in more detail. Under the Ptolemies, the Delta reached an ancient peak of production with 16,000 sq km of its 22,000 sq km under cultivation. This value exceeded the 11,300 sq km farmed in the entire rest of the Nile Valley and Fayum at that time. Today the cultivated lands of the Delta are about equal in area to those of all other regions, since the latter have been expanded by various reclamation projects and by multiple cropping. (When a field produces two crops a year it is counted twice toward area under cultivation.)

Classical writers such as Herodotus (around 450 BCE), Strabo (25 BCE) and Ptolemy (150 CE) described the presence of seven Nile branches (Fig. 11.1). Ranging from west to east these branches were the Canopic, Rosetta, Sebennitic, Damietta, Mendesian, Tanitic, and Pelusiac. By around 900 CE, however, only the two modern branches (Rosetta and Damietta) were functioning. Many sites of archaeological importance were located on the now defunct branches. Whenever the river changed course, it left these cities without a source of water for irrigation and navigation, and they were generally abandoned. At first they were targets of builders wanting to reuse their cut stone blocks, since the Delta lacks sources of stone, and old mud bricks were sought by farmers to fertilize their fields. Eventually, however, the ruins were covered by silt, often making their exact whereabouts a mystery.

An interesting example of a city destroyed by the forces of nature is that of Tennis (not the same as ancient Tanis). In the early centuries CE the region now submerged by Lake Manzala was arable land, unrivaled for its climate, fertility, and wealth. Its gardens, palm groves, vineyards, and grain fields were watered by branches of the Nile. In 365, a gigantic sea wave from the Mediterranean broke through the barrier sand dunes protecting the land on the north and flooded the low-lying land to form a brackish lake. This sea wave resulted from an earthquake centered on the Hellenic Arc, an east–west zone underlying Crete where the African Plate is being subducted beneath the Eurasian Plate.

The lake waters covered a number of towns—many of them renowned for weaving fine fabrics. Tennis became an island that was reached by an old branch of the Nile (probably the former Tanitic branch). Even as late as the tenth century CE, Tennis had many ancient monuments, among which were 160 mosques, 72 churches, and 36 baths. It had 10,000 shops and 50,000 male inhabitants. The city grew nothing and was entirely dependent on trade, especially with the Levant. Although it was surrounded by brackish lake water for most of the year, the annual Nile flood washed away the salt water and allowed the cisterns below the city to be filled with fresh water for the coming year. Today there is practically nothing left to see at Tennis, although archaeological excavations are ongoing.

The geology of the Delta has always affected the coastal cities and Egypt's foreign trade. The Mediterranean coast is "a dangerous low-lying coast [with] shallows far out to sea" (Pryor 1988:22). In fact, the depth is less than two meters as much as two kilometers out to sea. Large trading ships often had to transfer their cargo to lighters instead of coming into port. In 1249 CE, during the Seventh Crusade, the large troop transports of King Louis IX of France, who captured Damietta, could not approach closer than about 15 km to shore. And in 1836 CE, Frenchmen attempting to carry the Luxor Obelisk to Paris had to wait for the annual Nile flood to carry them across the bar at the Rosetta mouth. The shallow coast was made more treacherous by the prevailing north or northwest winds that could drive ships onto shore in storms. The famous lighthouse at Alexandria was one of a series of lighthouses built along the Mediterranean coast to help navigators.

The Wadi Tumilat and Sea-to-Sea Canals

Northeast of Cairo, at the far southeastern edge of the Delta, is an elongated depression called the Wadi Tumilat. It extends about 52 km from east to west and averages 7 km in width. Because it seems to branch from the main channel of the Nile near the Delta apex, it appears to be another defunct distributary. But analysis of the sediments in the wadi shows that it may at times have carried water toward the Nile instead of away from it.

When sea levels were higher than today, Nile water flowed eastward through the wadi toward an outlet near Ismailiya on Lake Timsah and perhaps south to the Red Sea. When sea levels dropped below the present level and the Nile was down-cutting its channels across the Delta, water drained from the Red Sea Mountains and flowed westward through Wadi Tumilat into the main Nile channel. This pattern reversed itself several times as sea levels rose and fell.

The Wadi Tumilat may be the oldest branch of the Nile, and in the early evolution of the Nile it diverted most of the river's sediments to the east. Later it became a natural route between the Nile Valley and the Gulf of Suez and Sinai to the east for caravans, since it cuts between high impassable hills and deserts to each side, and its low-lying areas offered water at or near the surface.

Many Egyptians kings must have wanted to find a better route between the Nile Valley and Sinai or the Red Sea than the overland routes through the Eastern Desert. (These overland routes are described in the chapter on The Eastern Desert.) The depression of the Wadi Tumilat leading toward Lake Timsah, the Bitter Lakes, and marshes north of the Gulf of Suez may have seemed the ideal course for a water route. In fact, it has been utilized in modern times as the route of a highway, a railroad, and the freshwater Ismailiya Canal.

Pharaohs of the New or even the Middle Kingdom may have wanted to excavate a ship canal through the Wadi Tumilat, but Necho II in the Twenty-sixth Dynasty (r. 610–595 BCE) is credited with actually starting one. According to Herodotus, Necho abandoned the project after 120,000 workers died—leaving the canal to be completed by the Persian conqueror Darius (r. 521–486 BCE). Darius, who wanted a sea route between the eastern Mediterranean and his distant Indian territory, recorded this building feat on a large stela.

Excavations undertaken at Tell el-Maskhuta (see Holladay, 1982) tell a slightly different story. They have revealed the remains of a city that guarded the eastern end of the canal. The city was built in the late seventh century BCE and grew during subsequent centuries when the canal was busy with trade goods from India and Arabia heading for the Mediterranean markets. The city continued in existence until Roman times, when it became a garrison town—more concerned with the eastern defenses than with trade.

While Herodotus was wrong about the date of the canal's construction, he was probably correct in his report that it branched off the Pelusiac branch above the city of Bubastis (near modern Zagazig) and was wide enough for two rowed triremes to pass (about 40 m). This canal was still usable, or was renovated, during the Ptolemaic period, but may have seen a decline in use during the first centuries BCE and CE. The emperor Trajan (r. 98–117 CE) ordered the canal improved to facilitate his conquest of Arabia. Soon, however, the use of larger ships and a dependence on sail caused this shorter route to be abandoned entirely as it was nearly impossible to sail into the Gulf of Suez against its strong north winds. Instead the Roman ships engaged in the Indian trade docked in ports along the Red Sea coast, and their goods were transported overland to the Nile Valley.

Like all branches of the Nile, the canal in the Wadi Tumilat was probably subject to silting and filling with wind-blown sand and thus required constant maintenance. It must have been unusable at the time of the Muslim conquest, but the Muslim general Amr apparently had the canal dug out again so that he could ship grain directly to Arabia.

The Delta in Modern Times

Today the Nile has only two distributaries that reach the Mediterranean Sea, at Rosetta and Damietta (Fig. 11.2). The Rosetta branch draws off 70 to 80 percent of the water, while the Damietta branch gets the smaller portion. The Damietta branch has a lower gradient and would silt up entirely if it were not maintained by dredging. Very little fresh water reaches the sea at the mouths of these branches: the majority of the water is diverted into thousands of irrigation canals, totaling more than 10,000 km in length. (The majority of these are not shown on Fig. 11.2). After flowing across the cultivated land, the water is collected in drainage ditches and returned to the original Nile branches, much reduced in volume but carrying large quantities of fertilizers and industrial wastes. A great deal of the diverted water reports to the lakes or lagoons rather than back to the river. As a result the water quality in the lagoons is falling, causing a reduction in fish productivity and a change in the ecosystem on which many native and migrating birds depend.

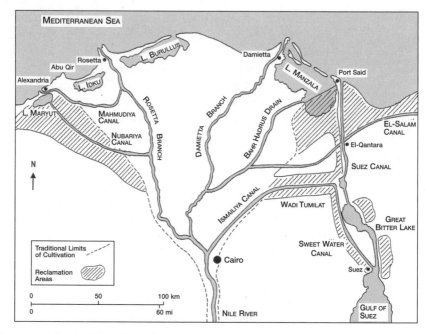

Fig. 11.2: The Nile Delta in modern times, showing the two functioning Nile distributaries. The Delta is also completely laced with irrigation canals, which are not shown on this map. Many feeder canals or drains follow the courses of ancient Nile distributaries. L = Lake.

Before the flow of Nile water and sediments was altered by the construction of dams, an average of 110 million tons per year of sediments was measured at Wadi Halfa. Of these, 58 million tons remained in the river by the time the water reached Cairo, and of this about 25 million tons were delivered to the Mediterranean, while the rest was deposited on the surface of the Delta. Deposition in the Delta, where perennial irrigation with no annual flooding has been in effect since the beginning of the twentieth century, is only 0.6 cm/century, but may have been as high as 12.5 cm/century before any dams were built. The Delta was growing along its coastal margins at a rate of about 1 cm/century, primarily in the regions around Rosetta and Damietta.

Although the High Dam now traps all the sediments of the floodwaters, the river still picks up sediments from the floodplain and riverbed as it moves downstream. Most of this gets transferred to the Delta lands as the water flows through irrigation canals. The irrigation waters flowing off the fields upstream also carry an increased load of salts from fertilizers. This too is transferred to the fields in the Delta, which are becoming waterlogged and salty. This problem is exacerbated by the use of perennial irrigation on two or even three crops per year. Two solutions are being used to solve the problems and restore the legendary fertility of the region. These include more efficient methods, such as drip irrigation to reduce the amount of irrigation water applied to the fields and the installation of drainage systems to draw off the excess water.

Some authors claim that the Delta coastline is eroding as a result of the decline in sediments: that is, wave action is no longer counterbalanced by fresh supplies of sediment. Some also claim that the Delta is subsiding. Rushdi Said (1993) disputes both these claims, noting that in contrast to deltas composed almost entirely of clay or silt that can undergo significant compaction, the Nile Delta has only a thin silt layer supported on many meters of noncompactible rocks, gravel, and sand—the deposits of the **Prenile** River.

Said attributes the undeniable submergence of the Delta margin to gradually rising sea levels. He notes that in the seventeenth to nineteenth centuries CE, Turkish forts were built on a coastline that is now about up to 8 km into the sea. Since then the sea level has risen, submerging these structures as it has Ptolemaic buildings in Alexandria. According to Karl Butzer (1976), the sea levels have fluctuated up and down by several meters in the past five thousand years. Because the gradient of the Delta is so small, an additional rise in sea level, as predicted by advocates of global warming, would inundate large portions of the Delta coastline. This issue is still one of great concern and active study, however. Regardless of the natural factors at work, human interventions are making most problems worse, and it is not clear if any solutions are possible, given the rapidly expanding populations and the desire for development.

Fortunately, the deteriorating surface water is not the only source for domestic supply. The porous Prenile River's sand and gravel layer serves as an excellent reservoir of fresh water. It is nearly 100 m thick, with an estimated capacity of 300 billion cubic meters, or the equivalent of three and one-half years of total Nile flow. This aquifer is replenished by seepage from the Nile and irrigation canals. As the water passes into and through the sand, some of the impurities are filtered out. This reservoir has already been heavily tapped, however, with the consequence that seawater has penetrated into the reservoir from the north and now extends at least 35 km inland.

Since the Aswan High Dam was completed in 1970, a number of desert development schemes have been implemented or are proposed to extend the cultivated area of the Delta to both the east and the west. This has involved the extension of existing canals or the excavation of new ones to provide water to the margins (see Fig. 11.2). The Nubariya Canal provides irrigation water to regions west of the traditional Delta as far north as Lake Maryut. On the east, the existing Ismailiya Canal is being extended to irrigate 310 sq km on the east side of the Suez Canal. And a mammoth new canal known as El-Salam (Peace) Canal has been constructed eastward from the Damietta branch. The fresh water from the Damietta branch will be combined with runoff carried by the Bahr Hadrus Drain. This canal will eventually stretch 240 km and provide water to nearly 2,500 sq km of land south of Lake Manzala and along the Mediterranean coast on the east side of the Suez Canal. It is interesting to note that when the Salam Canal was proposed, objections were raised that Egypt was moving Nile water outside of its 'basin' in contravention of International Treaty. Citing the old Pelusiac branch of the Nile, which opened east of Port Said, the Egyptians claimed that the region was historically part of the Nile basin.

Alexandria and the Northwest Coast

The northwestern coast of Egypt, extending from Abu Qir to Alexandria and west to the Libyan border, has a different geological makeup from that of the Delta's Mediterranean coastline. In the Delta east of Abu Qir the coast is lined with sand dunes composed of quartz sand and silt carried to the Mediterranean by various Nile distributaries. West to east currents distribute this sediment into long sand bars enclosing lagoons. The northwest wind has blown sand from the bars onto the shore south of the lagoons to form dunes.

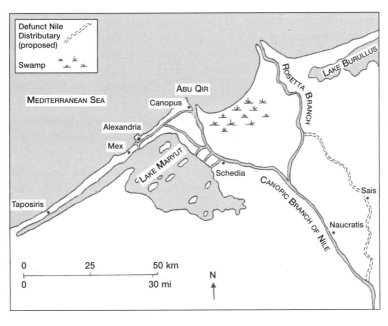

Fig. 12.1: Map of the northwestern Delta and the region around ancient Alexandria.

West of Abu Qir the coastal plain is interrupted by a series of **oolitic** limestone ridges. Two of these ridges run nearly continuously and parallel to the coast from Abu Qir to Salum on the Libyan border. The more northerly ridge has an elevation of about 10 m above sea level, while the one south of it is about 25 m in elevation. Six or perhaps more ridges run in a southwesterly direction from approximately south of Alexandria to just south of el-Alamein—a distance of about 100 km. Their crest lines are less uniform in height, but the ridges generally increase in elevation as one proceeds south in the series until heights of more than 100 m above sea level are encountered. Another limestone ridge appears about one kilometer north of the coastline where it appears as low islands or barely submerged rocks. Several more ridges may lie further offshore, adding their submarine hazards to the already dangerously shallow shore. Between the long limestone ridges are depressions. Those near the shore may lie low enough to be filled with salt marshes or freshwater lakes (like Lake Maryut). The depressions farther back from the shore may be cultivated.

Geologists have long debated the origin of these limestone ridges. Writers in the early twentieth century proposed that ridges formed from wind-deposited dunes composed of sand-sized grains of calcium carbonate derived from seashells and other marine debris. They suggested that acidified rainwater dissolved some of the calcium carbonate, and then when the water evaporated the mineral precipitated, cementing the sand grains together.

Since the 1950s, however, a different explanation has prevailed as a result of more careful study of the microstructure of the oolitic grains and the chemical composition of the cement. These studies suggested that each ridge was formed as an offshore bar during a period when sea levels were different from today. The depressions represent ancient lagoons between an offshore bar and the coast.

Fossils present in various ridges show that all were formed during the Pleistocene Epoch, and that the ridges decrease in age as one approaches the modern coastline. That is, the oldest ridges lie farther south. For these ridges to be formed so far inland (by modern standards), sea levels must have been much higher than at present or the land must have been elevated subsequent to their formation. Ridges now submerged must have been formed when the sea level was somewhat lower.

The Pleistocene is also called the Ice Age. During this geological Epoch, the repeated formation and retreat of glaciers over a several-hundred-thousand-year cycle resulted in the drop and rise of world sea levels by hundreds of meters. While all the ridges originated as offshore bars, some were eventually exposed above the sea, and then they were augmented by wind-blown sand that was cemented by rainwater and precipitation just as earlier theorists predicted. This explains the variation in height of a ridge along its length. The presence of these limestone ridges was to be important to the inhabitants of this coastline.

Along the entire north coast of the Delta there are no natural deepwater harbors. After Alexander the Great conquered Egypt in 332 BCE, he traveled to the Siwa Oasis to consult the oracle there. On the way he surveyed the site of Alexandria and selected it for a new port city. The location had several advantages compared to other Delta locations: it had a firm bedrock foundation that was not flooded during the annual Nile inundation, a local supply of limestone for building, and a protected anchorage for ships behind the limestone island of Pharos. Fresh water was available in the nearby Canopic branch of the Nile. Alexander died in Babylon in 323 BCE, and the construction of his namesake city was carried out by his Ptolemaic successors.

The Ptolemies were more than simply kings of Egypt; they had ambitions for a Mediterranean empire. For this they needed an enormous navy and facilities to sustain it. They also needed to develop the resources of the country in order to build up its industries and agriculture to generate products to trade. They expanded trade in luxury goods with the Far East, and the entire economy was put under the control of a highly organized and efficient bureaucracy.

The city of Alexandria became the capital of Egypt and continued in this role until 642 CE. It replaced an earlier Greek colony at Naucratis as the major port for European markets. Naucratis, a town on the Canopic branch of the Nile, had been founded by veteran Greek mercenaries during the Twenty-sixth Dynasty and had been the chief trading city of the Delta. Like other Delta cities such as Tanis and Pi-Ramesses that served as Mediterranean seaports, it lay inland, not on the coast. This was probably a consequence of the low-lying coastline and marshy conditions farther north. Sais, the capital during the Twenty-sixth Dynasty, was almost certainly built on a branch of the Nile, but in about 25 BCE Strabo reported that it lay two *schoeni* from the river, another victim of the capricious changes of course characteristic of the Nile distributaries. The Ptolemies maintained the town of Pelusium, on the

131

Pelusiac branch, as the frontier town at which goods from Syria and the east paid custom duties.

Two excellent harbors were created at Alexandria by joining Pharos Island (an emergent section of an offshore oolitic limestone ridge) to the mainland with a stone causeway about 1.5 km long. This causeway provided access to the island and carried an aqueduct; it also had several arches in it to permit ships to pass between the two harbors. The harbor to the west of the causeway was for commercial shipping, the one to the east for the navy and royal yachts. The famous lighthouse was erected on an islet at the east end of Pharos Island in 279 BCE. According to most authors, it was over 100 m high and equipped with a heliograph by day and fire at night. This lighthouse and others along the Mediterranean coast were necessary for the safety of ships attempting to make landfall on the shallow Egyptian coast with north winds at their backs. It was neglected after the Arab conquest and finally succumbed to an earthquake. The Egyptians used some of its blocks to build Fort Qaitbay on the same site in 1480 CE.

The former hazards of the coast are being revealed today as marine archaeologists explore a number of shipwrecks dating to many periods. They have also found the remains of buildings and sculptures lying just offshore near the site of the former lighthouse. Some of the sculptures bear inscriptions of earlier Egyptian rulers, indicating that the Greeks and Romans moved many items such as obelisks, sphinxes, and columns from Heliopolis and Memphis to adorn the city of Alexandria.

Alexandria is not, and never was, on one of the Nile distributaries. The closest branch in 300 BCE was the Canopic branch, opening just to the east of Abu Qir at the east end of the limestone ridge on which Alexandria was built. A 20-km canal was dug along the limestone ridge from Alexandria to the Canopic branch to bring fresh water to the city. Water from the canal, which ran outside the southern city walls, was distributed into the city by several subsidiary channels. The canal itself finally terminated in the Western Harbor.

Several more canals were dug from the Canopic branch into Lake Maryut (called Lake Mareotis by Classical writers), which lay in the depression between the Alexandrian ridge and the one south of it. These canals provided shipping access to the rest of Egypt via the Nile. This inner shipping basin on Lake Maryut was even busier than the outer harbor that received Mediterranean vessels, as the Ptolemies became entrepreneurs on an interna-

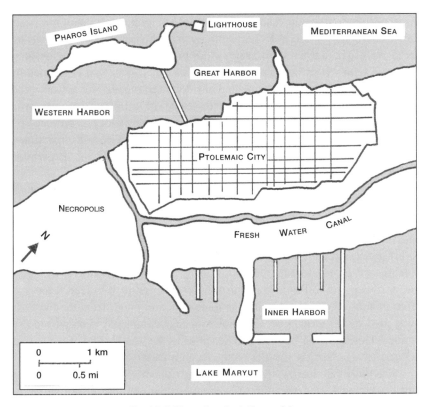

Fig. 12.2: Map of ancient Alexandria.

tional scale. Nile ships brought Egyptian products such as grain, linen, and papyrus for export to Mediterranean countries and ingredients for perfumes that were manufactured in Alexandria. They also completed the transfer of exotic goods originating in central Africa, Arabia, and India from cities in Upper Egypt to which caravans conveyed the goods from Red Sea ports. Some goods also came from the Red Sea via a restored canal in the Wadi Tumilat. The northern shore of Lake Maryut was lined with quays and warehouses. A town called Schedia (after *schedia,* a pontoon bridge, located there) was built on the Canopic branch to collect tolls and regulate traffic between Alexandria and the rest of Egypt.

Ruins of former towns and numerous agricultural estates are located on both the northern and southern former shorelines of Lake Maryut. In

Greco-Roman times the lake was about 100 km long, but today it has been reduced to only 50 km in length, with much of the surface now saltpans or reed beds. The agricultural estates were primarily involved in wine production, with vast areas planted in vineyards. Their wine was consumed in Alexandria and shipped throughout the Mediterranean. The towns were involved in the transshipment of goods from within Egypt; they also produced amphoras and other items connected with the trade. The presence of these estates is believed to indicate that the climate in Greco-Roman times was wetter than it is today, when the coast receives only modest winter rains, about 190 mm per year.

Earlier chapters of this book described the efforts of the Ptolemies to increase the amount of land under cultivation in the Fayum and the Delta. Their objective was to grow wheat for export to Mediterranean cities such as Athens. When the Romans gained control of Egypt, they too treated it as a breadbox and entrepôt—but with Rome as the destination for wheat and luxuries. By one estimate, 150,000 tons of wheat were shipped annually from Alexandria to Rome in Augustus's reign, or enough to feed Rome for four months. Under Roman rule the trade in luxury goods from the Far East was also greatly expanded. Emperor Augustus completely reorganized the trade routes, utilizing Egypt's Red Sea ports as the western end of the Indian Ocean voyage (see the chapter on the Eastern Desert for more details).

The Ptolemaic freshwater canal from the Canopic branch and Alexandria itself were neglected after the Arab conquest, but not before silt carried by the canal into the harbors had converted the causeway between the harbors into an isthmus. In about the ninth century CE, the Canopic branch dried up, and the Rosetta branch became the westernmost distributary. As Alexandria declined, Rosetta and Damietta became the chief ports involved in Mediterranean trade. In 1820 CE Muhammad Ali sought to restore Alexandria as an international port and ordered the 70-km Mahmudiya Canal dug from the Western Harbor to the Rosetta branch of the Nile. Its route through the city was nearly the same as the old Ptolemaic canal. Like that earlier canal, it supplied the city with fresh water and permitted ships to sail from Alexandria to Cairo. Oceangoing ships could not use this canal, however, so goods and passengers arriving at Alexandria had to transfer to smaller vessels.

Most of the splendid old Alexandrian landmarks such as the museum, library, temples, and palaces are gone, but many of the Greco-Roman cemeteries and catacombs still exist. Egyptians living in the Nile Valley construct-ed their cemeteries on the desert margin above the floodplain, or dug rock-cut tombs into the limestone cliffs. But burials presented a problem in towns built on the alluvial soil of the Delta. In Alexandria, the Macedonian Ptolemies and later the Romans excavated a large number of elaborate underground cham-bers, similar to those they had created in Greece and Italy, in the soft oolitic limestone. Carved and painted decorations in these tombs show an intriguing combination of Greek and Egyptian motifs.

The Kom al-Shuqafa catacomb, one of the largest, dates to the second cen-tury CE and was used for several centuries. Here a spiral stairway, colonnad-ed halls, a banquet room, and numerous individual and family burial cham-bers have been carved into the oolitic limestone bedrock. This catacomb is open to the public, whereas many of the more recently discovered ones are not. In some cases the recent discoveries were made during archaeological salvage investigations, after which the tombs were destroyed to make way for modern constructions. The Ptolemaic cemeteries lay outside the city walls, but modern Alexandria has expanded to cover these areas.

The Ptolemaic level of the city lies nearly 10 m below the current street level. Some authors attribute this to subsidence of the city, others think that sea level rise is responsible since remains of Greco-Roman cities around the Mediterranean are similarly submerged. Earthquakes were responsible for some destruction of the ancient structures: at least twenty-two were recorded between 320 BCE and 1300 CE. Many of these earthquakes origi-nated in the Hellenic Arc lying 900 km northwest of Egypt. Such a quake in 365 CE caused a sea wave that deluged Alexandria, stranding boats on shore; 50,000 houses were flooded and 5,000 people drowned, according to ancient reports.

Another type of structure that was excavated into the limestone bedrock was the cistern. Dozens of cisterns have been discovered lying beneath the modern city. From old reports, it seems there may have been thousands. These were dug to provide a place to store the excess fresh water arriving in the canal during the Nile flood. The cistern of el-Nabih is open to the public. It consists of a vast pit with a three-story colonnade supported by long rows of columns, mostly reused from the Classical period. The construction of these storage vaults began with the Ptolemies and continued under the Romans. The cisterns seemed to be fairly watertight, but the silt that settled out of the flood-waters had to be cleaned from the bottom of the cistern each year or it would fester and ruin the next year's supply. Attempts were also made to prevent the earliest, most silt-ladened floodwaters from entering the cisterns by the use of regulators on the subsidiary channels. *Saqyas* were used to raise the water from the cisterns.

Oolitic limestone was quarried from the coastal ridge lying between the sea and Lake Maryut at the coastal town of el-Mex. It was used by Greeks and Romans living in the Delta as building stone, since it was the only such stone available in the vicinity. It was also used in the 1860s to construct breakwaters at Port Said, the city built at the north end of the Suez Canal. Quarrying of this limestone continues today at several sites that can be seen from the highway between Alexandria and Mersa Matruh. This is the source of the small snow-white blocks used across northern Egypt to build houses, walls, and other urban structures. Because the limestone is so friable and chalky, cement and plaster do not adhere to it well.

The Western Desert

The Western Desert includes all the land in Egypt west of the Nile; this is two-thirds of the entire country, or more than 680,000 sq km. It is the eastern end of the Sahara, and is sometimes referred to as the Libyan Desert. Much of this vast expanse is rather featureless, with rolling stone plateaus, sand sheets, and fields of sand dunes. There is some relief, however, in the form of high cliffs, isolated hills, and depressions within the surrounding higher plateaus.

Although the depressions occupy only a small fraction of the total area, they are important as sites of permanent habitation. The major depressions are shown on Fig. 13.1, and their areas and depths are listed in Table 13.1 for purposes of comparison. Some depressions are hard to measure because they fade out into the desert without a clear boundary. Many other smaller, uninhabited depressions exist in addition to these.

Table 13.1: Areas and depths of depressions in the Western Desert

Depression	Area (sq km)	Lowest point (m relative to sea level)
Bahariya	1,800	+113
Dakhla	410	+100
Farafra	10,000	+100
Fayum	1,700	−53
Kharga	3,000	−18
Qattara	19,500	−133
Siwa	750	−17
Wadi Natrun	100	+23

Almost all the depressions have a steep escarpment to the north while their floors meet the desert surface on the south. The depressions' northern rims are **questas**, in which harder limestone layers form erosion-resistant caps above the softer layers of shale and sandstone. The Dakhla and Kharga questas are found at the boundary between Nubian sandstone in the south and the Cretaceous shale and Paleocene chalk in the north. The Farafra and Bahariya depressions lie at the Cretaceous–early Tertiary limestones boundary, and the Siwa and Qattara depressions lie along an Eocene–Miocene limestone boundary (refer to Fig. 2.2). There are probably two reasons for this pattern: the layers of rock all tilt slightly to the north, and the erosion that created the questas moved from south to north, as will be explained below.

Formation of the Depressions

Geologists have debated for years about the processes that created these depressions. Some believe that water running above or below ground was the most important agent in carving the depressions. In this scenario, rainwater percolated through the fractured layers of limestone, dissolving the calcium carbonate and creating caves. The roofs of these caverns eventually collapsed and formed sinkholes (see Fig. 9.4). Rainfall, rock falls, and wind ablation then enlarged the cavity. Other authors argue that running water should have left evidence in the form of drainage channels, and that no such channels can be seen on the nearly flat surface of the desert. As a result, these authors have argued for the effects of wind alone. The timing of the depressions' formation has also been controversial, with estimates ranging from the Eocene to the recent past. Since the climate varied greatly within this time range, the questions of mechanism of formation and timing are intimately related.

A new hypothesis that answers questions about both timing and mechanism has been proposed on the basis of data obtained by space satellites. Radar has detected the remains of old drainage channels (the so-called **radar rivers**) that are now buried by sediments and sand to such an extent that they are invisible to observers on the ground. Using this knowledge, Issawi and McCauley (1992) proposed that the depressions were formed in conjunction with two ancient drainage systems that operated during the Tertiary Period over a period from 40 to 6 million years ago. These drainage systems not only formed the depressions, they removed hundreds of meters of the rock layers that formerly covered huge areas of the southern Western Desert.

Formation of the Depressions

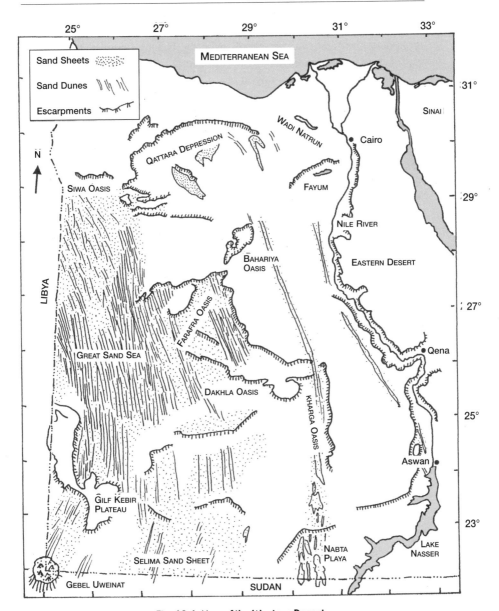

Fig. 13.1: Map of the Western Desert.

We recall that Egypt was submerged by the Tethys Sea intermittently throughout the Paleozoic, Mesozoic, and early Tertiary Periods. During phases of submergence, many layers of sedimentary rocks were formed one atop another—from oldest to youngest these layers included the Nubian sandstones, the Cretaceous limestones and shales, and the Eocene limestones (see Chapter 3). The land of Egypt was eventually lifted above sea level, but the rise was a gradual one, so that the north part of the country remained submerged and continued to accumulate sediments longer than the southern part. The uplift began in the southwestern corner of Egypt—the site of Gebel Uweinat and the Gilf Kebir Plateau—tilting the rock layers toward the north and producing cracks and joints in the surface rocks.

By the beginning of the Oligocene Epoch, or 37 million years ago, the sea had receded northward to the latitude of Siwa and Fayum, exposing a vast plateau of limestone over the southern part of the country. A humid climate produced intense precipitation and rapid weathering of the exposed rocks. Rainwater also seeped into the jointed limestone, forming underground channels and caverns. Runoff from the highlands of the Uweinat/Gilf area flowed north along the down-dipping strata. This gave rise to a river system that Issawi and McCauley called the Gilf River, which reached the receding Tethys Sea near Siwa (Fig. 13.2).

Although erosion on Gebel Uweinat rapidly carried away its upper layers of limestone and sandstone, uplift continued, thereby maintaining its elevation and ensuring that erosion would persist. Eventually the deep-lying Paleozoic sedimentary rocks and ancient **basement complex** were revealed. The Gilf Kebir plateau was not completely eliminated but remained as a massive block of Nubian sandstone over 1,000 m high. Huge wadis still penetrate this block and indicate the vigor of the former streams.

Additional tributaries to the Gilf River system arose later when the Red Sea Mountains began to be uplifted. Runoff and sediments flowed westward from the highlands and joined the Gilf River system at about the location of Kharga. This river system carried off the weathered Eocene and Upper Cretaceous limestones, exposing the layer of Nubian sandstone over the southern and southwestern parts of the country. Some uneroded, isolated limestone buttes still stand on the Nubian sandstone pavement. Tributaries to this river system carried off debris from the depressions of Kharga, Dakhla, and Farafra, which may have formed initially through the collapse of solid rock into underground caverns. As the Gilf River eroded

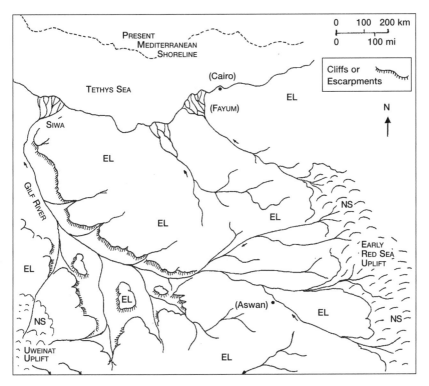

Fig. 13.2: The proposed Gilf River system of the Oligocene Epoch.
Arrows indicate direction of flow in the channels.
Rock layers at the surface: EL = Eocene limestone, NS = Nubian sandstone.
Reproduced with the kind permission of Dr. Bahay Issawi.

the rock layers in its catchment area, the gradient of the river decreased. This reduced its velocity, so that it began to drop portions of its sediment load and fill its channel with huge quantities of silt and sand. This material would eventually provide a source for the sand that would be blown by the northwest winds across the Western Desert to form the sand sheets and dunes we see today.

About 24 million years ago, as the old Gilf River system was losing its force, a powerful new drainage system developed to the east along a line close to the one followed by the modern Nile between Qena and Aswan. Fed by ample rainfall in a tropical climate, this system, which McCauley and Issawi

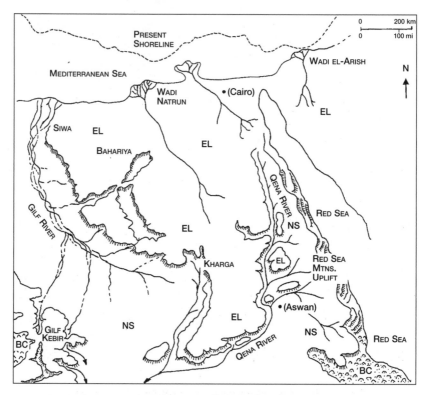

**Fig. 13.3: The proposed Qena River system of the early–Miocene Epoch.
Rock layers at the surface: EL = Eocene limestone,
NS = Nubian sandstone, BC = basement complex.
Reproduced with the kind permission of Dr. Bahay Issawi.**

call the Qena River, flowed south (Fig. 13.3). Evidently the high Eocene limestone plateau blocked a route to the north. Carrying runoff and sediments
from the still rapidly rising Red Sea Mountains, it began to cut into the limestone layers—previously eroded from the west by the Gilf River—from the
east and south. The result was the exposure of a layer of Nubian sandstone
over the southeastern part of the country. In some places—as at Aswan and
the southern Red Sea Mountains—the sandstone was completely eroded away
to expose rocks of the basement complex.

At Aswan the Qena River turned southwest and received tributaries from
what would later be known as Wadi el-Allaqi and Wadi Gabgaba. Near the

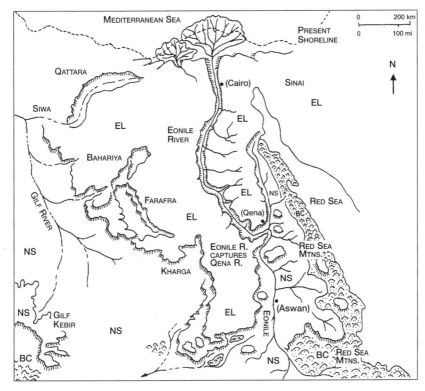

Fig. 13.4: The late-Miocene Eonile River system after it captured the Qena River system and diverted its flow to the north. R = River. Rock layers at the surface: EL = Eocene limestone, NS = Nubian sandstone, BC = basement complex. Reproduced with the kind permission of Dr. Bahay Issawi.

current Egyptian–Sudanese border a tributary flowing from the Kharga Depression may have joined it. The Qena River flowed into Sudan and may have continued on via the present Wadi Howar toward the Lake Chad basin.

Around six million years ago events occurred that led to the formation of the modern Nile system. One of these events was the desiccation of the Mediterranean Sea; another was the continued uplift of the Red Sea Mountains (see Chapter 3). At first, a small local stream, the **Eonile**, flowed north into the desiccated Mediterranean. But fed by heavy rains over the Red Sea Mountains and invigorated by the increase in gradient resulting from the lowered sea level, it rapidly began cutting a deep channel back southward

143

until it eventually intersected the Qena River system—perhaps the main river at Qena, or more likely it first met a Qena tributary flowing from the western plateau. This would explain the great bend in the River Nile at Qena. Because of the Nile's steeper gradient toward the north, where the base level was the falling Mediterranean Sea level, the Nile essentially captured the Qena's flow and turned what had been an independent, south-flowing river into part of a north-flowing one (Fig. 13.4). Tributaries of the Qena River, such as Wadis Shait, el-Kharit, Kalabsha, and el-Allaqi, became tributaries of the Eonile. Note that this new hypothesis does not contradict the account given in Chapter 3 for the formation of the River Nile system but augments it and explains additional features of that system.

One of the main strengths of the McCauley and Issawi hypothesis, in my opinion, is that it provides a coherent explanation for contemporaneous events in several parts of the country. The scenario described above explains how a landscape with a relatively uniform covering of three layers of rocks (basement complex, Nubian sandstone, limestones) became denuded of the limestone layer over its southern portions. Over the highest regions, the sandstone was also removed, revealing the ancient rocks of the basement complex. The remaining limestone forms a plateau across the northern part of the country with a tongue-like portion extending south between the Kharga depression and the west bank of the Nile. This plateau has an elevation of over 600 m above sea level just east of Esna on the Nile and slopes gradually to around 400 m in the escarpment east of Kharga.

The remains of the earlier river systems, the Gilf and the Qena, have almost been obliterated: their channels were filled with sediments deposited by ephemeral streams during the Pliocene and Pleistocene Epochs' pluvials and with wind-blown sand. The presence of the **radar rivers** has been confirmed by investigations on the ground. Once geologists knew where to look, they dug trenches into the desert that revealed the old sediments. The discovery of the ancient rivers confirms a suspicion of many geologists that the landscapes of arid lands cannot be explained by the processes operating under the current arid conditions alone; instead one must look back to processes operating during earlier, wetter periods for a complete understanding. This conclusion turns out to be relevant time after time in our study of Egypt.

While the bedrock of much of the Western Desert is Nubian sandstone, or in some locations limestone, this rock is often covered by a compact layer of sand called a sand sheet. Remnant outcroppings of the sandstone or limestone bedrock may protrude through the sand sheet. The sand sheet is usually covered by a layer of compacted gravel, known as desert pavement—the erosion-resistant remains of the limestone strata or wadi deposits. Wheeled vehicles can drive over this surface without getting stuck. Where the desert is topped by sand sheets, it has a gently rolling contour. In many places, however, sand dunes sit atop the sand sheet and form barriers to travel.

Water, Wind, and Sand

It is now clear that water—flowing both above ground and through subterranean channels dissolved in the limestone—was almost certainly the primary agent that produced the depressions in the Western Desert. This indicates that the climate was much warmer and wetter during the mid- to late Tertiary when they formed. Beginning with the Pleistocene, Egypt's climate became hyperarid, and the rivers of the Western Desert ceased to flow. During the occasional subsequent pluvials, however, **solution weathering** of the limestone may have helped to enlarge the depressions to their modern dimensions.

As we are now well aware, weathering also occurs when saline water percolates up from the sandstone. When the water evaporates from the surface of the rock it leaves salt crystals just below the rock surface; these crystals force loose fine flakes of stone that are easily **deflated** (removed) and carried by the wind. Winds can also remove the fine, dry silts left by drying lakes. This has helped to deepen those depressions in which lakes formed during past wet periods. Much of the surface of the desert is covered by an almost solid pavement of gravel left behind after all the loose fine material of former sediments was blown away.

Deflation is not effective on solid rock, but wind-blown sand can act as an effective scouring powder. A number of formations such as mushroom shapes and **yardangs** (streamlined oblongs that look like resting camels) have been created by sand carried up to a meter above the ground. Rocks have different resistances to wind erosion: shale and chalk (poorly cemented limestone) are easy to erode; Nubian sandstone, being loose-grained (also poorly cemented), is also very susceptible; well-cemented Eocene limestone is the most resistant and may form caps on **buttes** or other remnant blocks.

Wind has also been responsible for building the sand dunes found in many parts of the Western Desert. These dunes are of various shapes, including some crescents and stars, but most are longitudinal and run from north to south—a consequence of the northerly or northwesterly direction of the prevailing wind. The sand source, which must lie to the north of the dunes, has been attributed by some authors to the lower Miocene and Pliocene formations of the north coast, but more recent authors point to the layers of sand covering the river valley of the old Gilf River. The Gilf River must have carried huge quantities of sand eroded from the limestone and Nubian sandstones of the southern part of the country. Some of this sand reached the sea near Siwa, while much was deposited on the valley floor as the dying river became less powerful.

For 500 km along the Egyptian–Libyan border south of Siwa and corresponding to the location of the ancient Gilf River valley is a region called the Sand Sea. In the Sand Sea, there is a layer of sand topped by long parallel lines of dunes up to 150 m high. The dunes when viewed from a distance have no apparent organization but look like the choppy tops of waves on the ocean.

Between the Nile Valley and the depressions is a region with belts of crescent-shaped dunes running north-northwest to south-southeast. Each dune belt is only a few kilometers wide, but it may be hundreds of kilometers in length. It is nearly impossible for a vehicle to drive across one of these 30-m high dunes, although it can move easily on the hard gravel-paved surface between them. The great length of the formation, however, means that a detour around the dunes will stretch many kilometers.

Besides constituting a barrier to desert travel, the main irritation presented by sand dunes is their ability to cover up roads, rail lines, fields, and houses. Many sand dunes are in constant motion, and the study of their behavior shows that it is extremely complex. Planting vegetation can help to stabilize a dune and prevent its forward motion, but growth of vegetation requires more moisture than is available through most of the Western Desert. Nonetheless, one can see many small dunes with bushes sprouting out the tops. In these cases, a tiny underground source of water has permitted the growth of a bush that begins to trap wind-blown sand. As the sand pile grows, the bush must grow to keep from being smothered. Such dunes are termed phytogenic, or 'plant generated.'

Aquifers, Wells, and Oases

Today most of the Western Desert falls into the zone of hyper-aridity, with less than 5 mm of rainfall per year. In fact, over most of the desert no rain falls for years on end. Life in the desert is entirely dependent on ground water. Ground water is found in the layers of Nubian sandstone that occur across the Western Desert at various depths; impermeable layers of clay above the sandstone prevent the water from rising to the surface in most places. In depressions where more of the layers of rock above the Nubian sandstone strata have been removed, the ground water makes its way to the surface through springs or **artesian wells**. There we find the legendary desert **oases**. In other low areas in the southern section of the desert water can be obtained by digging shallow wells, called *birs*. But in a large area about 800 km by 300 km to the south of Siwa and west of Farafra, there is apparently no source of water near the surface.

Since none of the depressions has any external drainage, they act as collecting basins for any rain that may fall. During wet periods of the past, many of the depressions contained **playa** lakes that are now represented only by dry deposits. Siwa and Wadi Natrun still contain several lakes that are fed from artesian sources, while the Fayum's lake is maintained by water from the Nile. These lakes are generally too saline to use for consumption or agricultural purposes; instead they act as drains for irrigation water.

In order for a depression to support permanent habitation with at least some agriculture, fresh water must be available in a spring or by pumping from wells. Today only Siwa, Bahariya, Farafra, Dakhla, and Kharga have settlements. A number of other depressions were inhabited in the past but have been abandoned because the ground has become so saturated with salt that nothing—not even the slightly salt-tolerant date palm—can grow. This is a consequence of watering with slightly salty water: the roots of plants are adapted to prevent sodium absorption, so the salt stays in the soil, and as irrigation water evaporates from a field year after year, the salt concentration in the soil rises. Today, dying date palms can be seen in Siwa along with salty crusts on the fields and rimming the lakes. This suggests that this oasis too may become useless for agriculture if current irrigation practices (flooding of fields) persist.

It is now recognized that a huge underground store of water or **aquifer** is located beneath the entire Western Desert. It is contained within the porous Nubian sandstones that range in thickness from 500 m in the south to 2,500 m near Siwa. The total volume of water in the aquifer is estimated at 50,000 cubic km, an amount equivalent to about 600 years of River Nile discharge. It seems likely that there are in fact two layers of sandstone, and in places such as Kharga, artesian wells tap the upper layer, which has been exposed by erosion. The deeper layer that constitutes the main aquifer is sealed between impermeable shale above and the basement complex below.

The sources of this aquifer water are controversial and may include runoff from plateaus to the south and west, movement of water from aquifers to the south and west, local precipitation, and seepage from the Nile. None of these current sources appears to be adequate to account for such a large reserve, however, and so it is assumed that the water is the remnant of earlier local pluvials. This is confirmed by dating many ground water samples. Two ages of water have been identified: older than 20,000 years and less than 14,000 years. These two periods of ground water collection coincide very well with the dates of past wet periods as determined by other studies. Importantly, all the samples from New Valley sites (see below) showed ages in the 20,000 to 40,000 year range. Water is known to move in aquifers under the influence of gravity from areas of recharge to areas of discharge. The age of the water in the Nubian aquifer shows that recharge, while possible, occurs at such a slow rate that the water should be treated as a non-renewable resource.

Many schemes have been offered to use this ground water to expand agriculture and provide for new settlements. For example, President Nasser instituted the New Valley Project in the late 1950s to encourage development in a region encompassing the Oases of Bahariya, Farafra, Dakhla, and Kharga. But caution has been advised about tapping this resource, which cannot be renewed at anything like the rate is it used. The water will also become more costly as the aquifer is depleted and wells have to pump from deeper strata.

Human Habitation in the Western Desert

In spite of its aridity and seeming inhospitable nature, humans have both lived in and traveled through the Western Desert for hundreds of thousands of years. The availability of water has always limited human usage, however. Today, that means that people live permanently only in the few oases; in the future, wells tapping the Nubian Aquifer and transfers of water from the Nile may expand the areas that can be occupied.

Several lines of evidence show that during some periods of the past, wider occupation was possible because conditions in general were much wetter than today. This may have been the result of regional changes in the rainfall, whereby the rain bands now confined to lower latitudes moved north, or it could reflect a worldwide change in climate such as those that produced the Ice Ages. Evidence for prehistoric occupation extends back into the middle Pleistocene. This evidence has been the result of investigations that depended on both archaeology and geology.

Archaeologists, at first, concentrated their search for ancient humans in areas with clear signs of present or former water: the oases, wells, and playa lakes. The radar river discovery has broadened the scope of the investigation and brought under scrutiny new areas that attracted no attention before. Geologists have analyzed deposits to determine their ages and to reconstruct the climatic conditions during their formation. Artifacts recovered in the geological investigations as well as by archaeologists reveal the level of cultural development of the inhabitants and their means of subsistence.

As a result of a number of studies, a more complete picture of occupation of the Western Desert is beginning to emerge, but many details are lacking that would permit a complete correlation among the desert sites, between the Western Desert sites and the Nile Valley, and between Egypt and other Saharan countries. Such correlations are the ultimate goal of many research programs.

All results point to episodic occupation in the Western Desert, reflecting an oscillation in climate between arid or hyperarid and wetter conditions. The literature is too extensive to do more than provide a brief summary; furthermore, new data that may necessitate different interpretations are appearing at a rapid pace. The earliest occupational evidence—in the form of hand axes fashioned of flint or quartzite (siliceous sandstone)—dates to the mid-Pleistocene of more than 300,000 years ago, a time period archaeologists call the Lower **Paleolithic**. Many sites also provide evidence for occupation between 200,000 and 150,000 years ago. A period of aridity then followed. The next group of occupation dates, again from many locations, occurs in the Middle Paleolithic, within the range of 60,000 to 40,000 years before present. From 40,000 to 10,000 years ago, conditions were clearly very dry, and no evidence of occupation has been discovered in the Western Desert from that time. But the Holocene ushered in a Wet Phase that lasted from around 10,000 to 5,000 years before the present, with many Western Desert sites providing evidence of this stage of occupation.

A Closer Look at the Geology of Some Individual Regions

North of Siwa and the Qattara Depressions there is a plateau of limestone. It was formed during the Middle Miocene, when the rest of Egypt was above sea level. This limestone layer ranges from about 140 m thick in the south to 430 m in the north. Between the Miocene limestone plateau and the sea is a coastal plain of variable width composed of oolitic limestone ridges and sand dunes. The sand dunes are composed of grains made of broken seashells rather than quartz.

This coastal plain benefits from the Mediterranean climate, with its light winter rains of about 19 cm per year. Runoff seeps into the limestone or collects under the coastal sand dunes, rather than forming streams running toward the Mediterranean Sea. This ground water can be tapped by wells and stored in cisterns. The water table actually lies above sea level, and wells dug too deep penetrate into a layer of salt water. Interestingly, the water table under the dunes is affected by the height of the dunes: the water table rises higher under the dune than beneath the flat ground to either side of the dune. A well sunk between dunes will benefit, however, by drawing on the water stored beneath the dunes.

The Siwa Oasis occupies a depression with a wall of Miocene limestone as its northern border; the escarpment is 200 m high. The depression is bounded on the south by a plateau of Eocene limestone with an elevation of 500 m above sea level, but no scarp is visible because it is covered by sand dunes. The entire floor of the depression lies below sea level, reaching 17 m below sea level at its lowest point. Several permanent lakes occur in the depression. These were once larger and are now very saline. There are several hundred springs, but only about eighty of them can be used for drinking or irrigation; the others are too salty. The water supply could be greatly augmented by drawing on the Nubian aquifer, which is especially thick under Siwa.

One can see that the process of cliff erosion and retreat that formed the Siwa depression worked from south to north. The northern escarpment itself is much dissected by wadis running toward the south. Scattered around the depression and along the cliffs to the north are remnant hills of limestone, which served as sites for both towns and cemeteries. The cemetery atop Gebel el-Mawta provides a good vantage point over the entire Oasis. Tombs in this cemetery date from the Twenty-sixth Dynasty through the Roman period. Oasis residents used them as shelters during Second World War desert battles.

Building stone was readily available in Siwa and was used to build temples, but the houses are built of mud brick in the traditional style. The mud/sand material comes from lagoons, which are salty, and salt crystals help weld the sun-dried bricks together. When it rains, the salt dissolves and the bricks and houses disintegrate. Fortunately, there is essentially no rain in the Oasis. But every few decades, or even generations, a downpour occurs. The ancient town of Shali, set on a hill and surrounded by a high wall for protection from the roving nomads, was so badly damaged by a storm in the 1920s that it was abandoned. It now forms a very picturesque ruin.

The gigantic Qattara Depression is bordered on the west and north by a high escarpment of Miocene limestone; it is open to the east and south. Its average depth is 60 m below sea level, with a lowest point of 133 m below sea level. Its basin contains a salt-water lake. Among many schemes proposed for the Western Desert was one to dig a tunnel 75 km north to the Mediterranean so that seawater could flow into the Qattara Depression and power a generating station. A recent engineering evaluation of this proposal concluded that it is not economically feasible, given the current price of other forms of energy.

The Wadi Natrun depression is the most northerly and easterly of those in the Western Desert; it lies about 70 km northwest of Cairo. It is named after the mineral natron—a mixture of the compounds sodium carbonate and sodium bicarbonate. Today there are about a dozen shallow lakes strung along the floor. In earlier times, the lakes held more water and merged together, reducing their overall number. Although designated a wadi, there does not appear to have been any connection between this and the Nile Valley; instead it is another depression and like others in the Western Desert may have had the same origin. It may also have been the product of another ancient river system of which few traces remain.

Natron occurs in solution in the lake waters and as a layer of evaporites in the lake bottoms and around their borders. The presence of the natron seems to arise from waters—originating from both springs and perhaps the Nile—percolating through subterranean layers containing the salts that were laid down when the Wadi Natrun was a lagoon on the Mediterranean coast. The water carrying the salts into the lakes evaporates in the heat and leaves the salts behind. This process continues, but the rate of deposition is much slower than the rate at which the salts have been removed by human activity. The resource is exploited today just as it has been since pharaonic times, when it was used in the manufacture of glass and in mummification.

Bahariya is an elongated depression completely surrounded by cliffs. Its long axis runs northeast to southwest. Numerous isolated hills dot the floor, which lies an average of 100 m below the level of the surrounding desert plateau. These hills are dark colored because they are covered with various dark hued rocks such as **dolerite**, ferruginous or iron-bearing limestone, and **basalt**. The floor of the depression consists of a very deep layer of Nubian sandstone. The sandstone continues into the lower part of the surrounding escarpment, where it is topped by Eocene limestones and Oligocene basalts.

Bahariya has one of the few commercially significant deposits of iron ore in Egypt. The ore is found within the Eocene limestones; the method of formation is not agreed upon but probably relates to the same volcanic processes that produced the basalt in this area. This ore is used in the Helwan Iron and Steel Works, to which it is transported by a special rail line. Springs are numerous in this oasis, and in the past water could be reached by digging only 7 m. The Romans constructed a series of underground aqueducts to move the water. These can be located by the openings to the surface that allowed the channels to be cleaned.

The Farafra Depression has a triangular shape, with the apex at the north; it is bounded by high limestone cliffs (about 225 m high) to east and west and lower cliffs to the north. The upper vertical scarps are formed of Eocene limestone above Paleocene shales that weather into sloping piles of fine debris. The scarps retreat as the soft shale layers erode and undercut the limestone strata. Eventually, unsupported blocks of limestone break loose and fall to the desert floor. The floor of the depression is late Cretaceous limestone or **chalk** of a very pure white color.

Within the Farafra Oasis lies a region of great natural beauty called the White Desert. This area has become a popular destination for travelers and has been designated to become a national protectorate. Here a traveler with a four-wheel-drive vehicle can examine many formations that illustrate the geological processes outlined in this chapter and other parts of this book. Fantastic shapes have been sculpted out of the white Cretaceous limestone. This limestone was formed from the remains of microscopic marine organisms and contains very little cement, making it soft and friable. Erosion of this layer of the desert plateau has produced isolated snowy hills draped with brown mantles of weathering debris colored by iron oxides. As the hills are reduced still further, they become prey to the wind-blown sand—carried only a few meters above the ground. This sand scours the lower portion of the hill and creates the top-heavy 'mushrooms' that dot the plain. Certain components of the plateau's rocks are very erosion-resistant, however, and remain on the surface after many meters of limestone have been removed. These include flint nodules; pieces of gypsum; fossils such as sea urchins, mollusks, and coin-sized nummulites; as well as black ferruginous (iron-bearing) nodules. Other erosion-resistant formations are rounded concretions (called melons) and pyramid-shaped piles that were formed as reefs in the ancient seas.

The Dakhla depression is bounded on the north by a steep escarpment, with occasional wadis penetrating it. The base of the scarp is formed of Cretaceous shale topped by a solid layer of Paleocene limestone. The floor of the depression is Nubian sandstone. One isolated butte and another nearly isolated hill of limestone remind us of the erosion process moving from the south that removed the plateau and formed the depression.

Kharga is not really a separate depression but is a continuation of the one designated as Dakhla. While Dakhla has its long axis running east and west, Kharga runs north to south for about 185 km. It is bounded on the northwest, north, and east side by a magnificent escarpment over 350 m high that is capped by Paleocene and Eocene limestone. There is no definite boundary to the west and south; the width of the oasis varies from 20 to 80 km. Nubian sandstone forms the oasis floor; it is overlain by shale toward the east. This is the same stratum that is mined for phosphates in the Nile Valley north of Idfu. Phosphates are also mined here in Kharga. Kharga is part of the New Valley Project. Water for agriculture comes from deep wells drilled into the Nubian aquifer and a proposed canal from Lake Nasser. Some parts are threatened by the moving sand dunes. One of the distinctive features of the landscape is the yardang—a wind-sculpted stone that is shaped like an upside-down boat hull. Some have suggested that the shape of these natural sculptures inspired the ancient Egyptians to carve the Sphinx.

The Eastern Desert 14

The Eastern Desert is dominated by the Red Sea Mountains, which form a range extending from the latitude of Cairo southward to the Egyptian–Sudanese border and beyond. These mountains have been mentioned in several previous chapters, since their presence affected events in the Nile Valley and throughout Egypt. Their origin was described in Chapter 3 in connection with the formation of the Red Sea. And their important role as a source of runoff and sediments contributing to the three Tertiary river systems, the Gilf, the Qena, and the Nile, was described in the last chapter. As the mountains were repeatedly uplifted by tectonic forces, their streams were continually **rejuvenated**. The steep gradients and the abundant rainfall of earlier times combined to erode the rock layers severely. Thus both uplift and erosion are responsible for the dramatic scenery we see today.

There is a huge difference between the age of the rocks in the Red Sea Mountains and the age of the Red Sea Range itself. The igneous and metamorphic rocks found in the **basement complex** of eastern Egypt are tremendously old: 550 to 900 million years. They were formed during a series of collisions among various crustal plates. At that time these rocks were deeply submerged within the continental crust, and they probably lay beneath a high mountain range formed by the collisions. Gradually, the ancient mountain range was eroded away. Not later than the mid-Mesozoic Era, this eroded landmass was submerged beneath an invading Tethys Sea, and the layers of Nubian sandstone and Cretaceous and Eocene limestone formed atop it. After these layers were in place, the sea withdrew and rifting began to split apart this eastern edge of the African plate. The uplift of the area to form the Red Sea Mountains probably began around 35 million

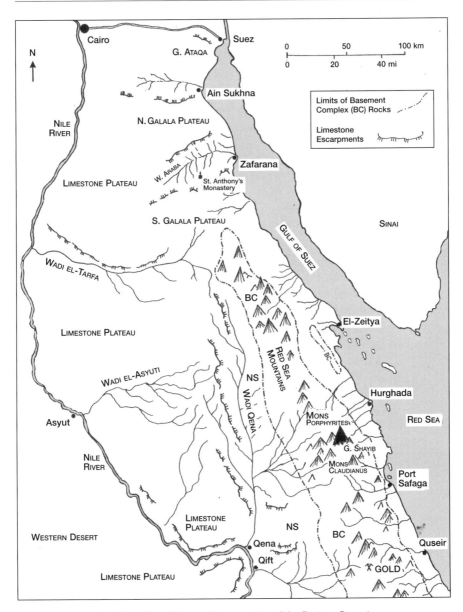

Fig. 14.1: Map of the northern section of the Eastern Desert.
Rock layers at surface: NS = Nubian sandstone, BC = Basement Complex. G = Gebel.

years ago in the Oligocene and continued into the Miocene and Pliocene. The mountains are thus relatively young by geological standards, but contain much older rocks.

Eocene limestone forms high plateaus over a wide expanse at the northern end of the Eastern Desert. In the middle and southern parts of the mountain range, however, the upper sedimentary rock layers have been sufficiently eroded to expose the rocks of the basement complex. These rocks, especially the coarse red granites, form the rugged and lofty peaks. The tallest pinnacle in the range, Gebel Shayib, is 2,184 m high, but many others are over 1,000 m.

Between the Nile Valley and the Red Sea Mountains are broad plateaus of sedimentary rocks. From Qena northward there are plateaus of Cretaceous and Eocene limestone with elevations of around 500 m. South of Qena the plateau consists of Nubian sandstone, with occasional uneroded remnants of limestone and shale forming peaks on the gently rolling sandstone surface.

The Eastern Desert is intensely dissected by deep wadis and ravines that either run eastward toward the Red Sea or westward toward the Nile Valley. Since the divide between the two drainage basins lies closer to the Red Sea than to the Nile, the wadis draining eastward are short and steep. The longer wadis of the westward drainage frequently coalesce and form a small number of major trunk wadis, including Wadis el-Tarfa, el-Asyuti, Qena, Abbad, Shait, el-Kharit, and el-Allaqi. (Fig. 14.1 shows the major wadis only. The wadis have been left off Fig. 14.2 for simplicity; some of these are shown on Fig. 7.1.)

We can begin our tour of the Eastern Desert at the Cairo–Suez highway, which marks the northern end of the Desert. This route to the Red Sea is more direct than the one through the Wadi Tumilat (described in the chapter on the Nile Delta) and was followed by mining expeditions heading to Sinai as well as by Christian and Muslim pilgrims. Leaving Cairo, the road ascends to a broad plateau of Eocene limestone. In places this bedrock is covered by a spread of sand and gravel deposited during the Oligocene while this area was still near sea level. This deposit can be viewed at the Petrified Forest on the north side of the road. This area was formerly mined for its sand and gravel, but it is now protected. The winds have swept away the fine sand leaving a pavement of gravel consisting mostly of flint nodules and chunks of petrified wood, as well as occasional tree trunks up to 10 m long. These trees were probably carried from regions farther to the south by the brisk streams running off the uplifted Red Sea hills. After they came to rest in shallow lagoons,

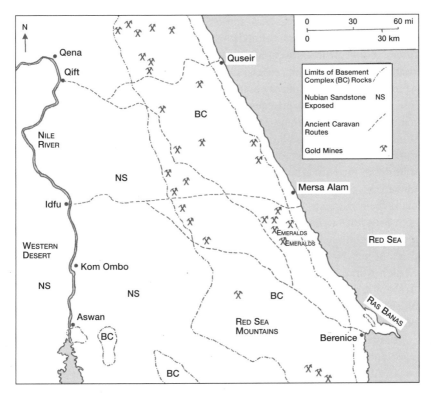

Fig. 14.2: Map of the southern section of the Eastern Desert, showing ancient caravan routes and gold mines. The many wadis draining to the Nile or the Red Sea are omitted for simplicity; some of these are shown on Fig. 7.1.

they were buried. Later, volcanic vents opened releasing hot water charged with silica. This silica 'petrified' the trees by replacing their organic compounds. Later, the Oligocene coastal deposits were uplifted to their present heights, and erosion revealed the petrified trees.

Two enormous Eocene limestone plateaus, called the Northern and Southern Galala plateaus, stretch across the northern part of the Eastern Desert. Their elevations are 1,274 m and 1,464 m above sea level respectively. Between these plateaus lies a great valley called the Wadi Araba, which has its mouth on the coast of the Gulf of Suez. Rocks of Cretaceous age and older are exposed at the base of the steep scarps on either side of the wadi and on the

157

valley floor, and their presence and orientation provide evidence about the geological processes that formed this valley. We have mentioned tectonic forces acting vertically to uplift a region, but these forces can also act horizontally to stretch or compress a region. Sometime after the Cretaceous rocks were deposited, and probably before the Eocene limestone formed (since the limestone strata in the plateaus appear horizontal when observed from the road to St. Antony's Monastery), this area was compressed in a northwest to southeast direction—perhaps related to the collision of the African and Eurasian plates. This produced a fold of uplifted rock running northeast to southwest; rock strata to the north and south of the fold were tilted upward toward the centerline of the fold. This folding cracked the rock layers and made them more susceptible to erosion. The top of the fold may have been above sea level or have received only a thin layer of Eocene limestone, while the areas north and south of it were deep basins that accumulated the hundreds of meters of rock now uplifted into the Galala plateaus. Subsequently erosion worked down through the layers of folded rocks forming the Wadi Araba and revealing on its floor tilted layers of Nubian sandstone and even older sedimentary rocks formed during the late Paleozoic Era over 300 million years ago.

South of the Galala Plateau, the elevation increases along the Red Sea coast and rocks of the basement complex begins to be revealed in rugged peaks. Along the Nile Valley, however, the limestone plateau, with elevations of 500 m, continues south to the latitude of Qena. This plateau is a continuation of the limestone plateau that lies west of the Nile south of Qena. The limestone plateau ends on the eastern side in a high escarpment overlooking the wide Wadi Qena, which separates the plateau from the mountainous region along the Red Sea coast.

Wadi Qena is unusual: its channel is nearly 350 km long and follows a course southward, almost exactly opposite that of the River Nile. Its course is on nearly the same meridian as the Nile south of Qena. This wadi has previously been identified (in Chapter 13) as the upper part of the Tertiary Qena River system. This river is postulated to have begun flowing toward the south around 24 million years ago. Then, around 6 million years ago, when the Nile was cutting its canyon southward during the late Miocene desiccation of the Mediterranean, its headwaters intercepted the Qena River (or one of its tributaries), and because of the steeper gradient back toward the north from Qena, the Qena River was captured (or **pirated**) by the north-flowing Nile. The course of Wadi Qena was cut in the stratum of easily-eroded Nubian limestone.

Variation in the reaction to erosion of different rock types determines the kinds of landscapes formed in this desert: limestones, which usually consist of alternating strata of dense limestone and softer shale and marl, erode into steep walled cliffs (dense limestone) alternating with sloping **scree** falls (shales). The limestones usually form caps over underlying slopes. Nubian sandstone is very loose-grained (poorly cemented) and erodes easily. Since it lacks layers of differing hardness, it erodes into a wide valley with a smooth sloping surface, except that **joints** may erode into deep vertical clefts. Granites erode into rounded hills, or where previously thoroughly-jointed, into knife-edged ridges and jagged peaks. Metamorphic schists erode into ser-rated peaks, while dikes of igneous material harder than their surroundings give rise to long protruding ridges.

The intense erosion of the Red Sea Mountains is proof that climates were formerly much wetter and precipitation over the mountains was heavy. Today, as in all of Egypt, the climate of the Eastern Desert is arid, but some rain falls on the mountains. This rain usually falls in the form of a single drenching downpour covering a small area, which runs off into a local wadi, converting it for a few hours or a few days into a roaring torrent. Nonetheless, most of the wadis are choked by coarse sediments that the inad-equate runoff can no longer carry past their mouths into the Nile. The runoff is trapped behind the debris and sinks into the rubble. This ground water encourages vegetation and supplies wells (*birs*), which generally need to be only 8 to 10 m deep. Water is also found occasionally in pools eroded out of the igneous rocks. The quantity of water in such a pool depends entirely on the recency of the rainfall in the area.

Human Occupation

In historical times, the wadis of the Eastern Desert no longer had perennial streams, but they still provided a reliable source of ground water in their debris-filled channels and a route among the high peaks. Men followed these routes from the Nile Valley into the mountains to mines and quarries; they fol-lowed the routes through the mountains to the Red Sea ports. Except for these purposes, the only inhabitants of the Eastern Desert were nomadic herders. The wadis lying south of Qena pass through a plateau of Nubian sandstone before they enter the Red Sea Mountains. The erosion of the sandstone has produced channels with easy gradients. Today, highways have been built along some of the ancient routes—from Qena to Port Safaga, from Qift to Quseir, and from Idfu to Mersa Alam—and these provide travelers with glimpses of magnificent scenery and the remains of human activities.

Much of the human activity in the Eastern Desert involved mining and quarrying. The igneous and metamorphic rocks of the Red Sea Mountains were the source of a number of rare and valuable stones. These included emeralds and amethysts for jewelry, amulets, and inlays. Steatite, a soft metamorphic rock, was used to make beads, amulets, and vases. Galena (lead sulfide) was used for eye paint. Diorite, various porphyries, greywacke (a type of sandstone), and other hard stones were valued for sculptures. A number of important minerals and metals are also found associated with igneous rocks: among these gold and copper ores were most sought after by the ancient Egyptians. Mineral deposits usually occur in or near fault lines, because faults provide a pathway along which mineral-rich fluids can rise through older rocks.

Gold was mined at many locations throughout the Eastern Desert and in the Nubian Desert southeast of Aswan (some of the most important sites are shown in Figures 5.2, 14.1, and 14.2). Modern geologists have identified 95 sources of gold in the mountains of the Eastern Desert, and nearly all of these had been discovered and worked by the ancient Egyptians. This suggests that they prospected their territory carefully and were skilled at recognizing exploitable resources. Gold was dug from open pits as well as underground mines. The process of separating the gold from the vein of quartz rock was incredibly laborious and involved hammering rocks to a powder, followed by washing with water to separate the grains of gold. Gold was used for jewelry, ritual objects, and gilding. It also provided a form of wealth that supported international trade and diplomacy.

The largest and richest gold deposit in Egypt, El Sid, was located along the route through the Wadi Hammamat, leading from Qift on the Nile to Quseir on the Red Sea coast. It was extensively exploited by the ancient Egyptians. As recently as 1944, mining resumed there, and 2,653 kg of fine gold was obtained in fifteen years. Qift became a prosperous city as a result of the traffic through the wadi. To the north of Bir Umm Fawakhir on the modern road are remains of stone huts used by late Roman or Byzantine gold miners. Ruins of watch towers and enclosures used by trade caravans can also be seen nearby. Many inscriptions—from prehistoric rock drawings to records of pharaonic mining expeditions, to Greco-Roman and modern graffiti—have been discovered along this route, documenting its importance and use in ancient times.

Wadi Hammamat was also the location of a quarry for siltstone (a hard, fine-grained, greenish or gray rock). Both siltstone and a slightly coarser form called greywacke sandstone were used for statuary, bowls, and sarcophagi. These quarries were worked throughout Egyptian history from Predynastic through Roman times. Another Roman quarry in the area yielded granodiorite, which was used for floor tiles and columns.

Many other useful metallic minerals have been identified in the Eastern Desert and other parts of the country, but most of the deposits are small or occur in concentrations that are too low to be commercially profitable today. This has been attributed to the fact that such minerals are usually found near the tops of igneous formations, and in the case of the Red Sea Mountains and Sinai, many hundreds of feet of this kind of rock have already been eroded away, taking their valuable deposits with them. Nevertheless, the number of mines and quarries worked by the ancient Egyptians, beginning in the earliest times through the Roman era, is amazing.

Two famous hard stone quarries lie north of the route between Qena and Port Safaga. Although the quarries are accessible only by rough roads, the visitor is rewarded with views of abandoned columns and other architectural elements. In addition to the mine sites, one can see ruins of workers' huts, a fortification, and a temple. Imperial Rome was the chief consumer of the attractive black-and-white quartz diorite **gneiss** from Mons Claudianus and purple andesite-dacite **porphyry** from Mons Porphyrites. The quarries were worked extensively during the reigns of Trajan (r. 98–117 CE) and Hadrian (r. 117–138) to provide columns and other elements for many buildings in Rome and throughout the empire. It is worth noting that coarse red granite is the most abundant rock in the Red Sea Mountains, but it was easier to quarry this at Aswan, where the blocks could be moved to the river easily for transport. The quarries in the Eastern Desert were often far from the River Nile or Red Sea coast, so transport of quarried blocks was difficult. Only desirable stones that were not available in a more convenient location were obtained there.

The stone quarried at Mons Porphyrites is truly unique: no other source of this particular rock is known anywhere else in the world. The location of this quarry was forgotten after the Roman period and was only rediscovered in 1823. This quarry produced the so-called imperial porphyry, which was used for royal sarcophagi and to panel the chamber in which princes were born—giving rise to the phrase 'born to the purple.' The rock has a deep reddish purple groundmass with large white or pink crystals. The Roman quarries lay high on the stony hills surrounding Wadi Abu Maamel (near Gebel Dokhan); blocks were lowered to the wadi floor down slipways. The blocks were dragged for 16 km along the winding wadi to a loading ramp. Natural boulders of purple porphyry can be seen among the deep layer of rocks on the wadi floor and these may have alerted the Romans to their source higher up. From the loading ramp, the stone was carried on wheeled carts 150 km to the Nile at Qena.

The Eastern Desert, like some parts of the Western Desert, attracted Coptic hermits. St. Anthony was one of the most famous—attracting so many followers that he had to take refuge in a cave. The Monastery of St. Anthony was founded in the fourth century after the monk's death in 356 CE. The monastery occupies a geologically interesting site: it lies on the south side of the immense Wadi Araba, against the scarp of the South Galala Plateau. Several springs debouch from the limestone cliffs and provide a reliable source of water in this dry wasteland. This water supply permits the present-day residents to cultivate a variety of fruits and vegetables in the monastery gardens. The same is true at the Monastery of St. Paul, slightly farther south along the Red Sea coast.

The Coastal Cities

The Red Sea Mountains are flanked to the east by a coastal plain composed of down-faulted crustal blocks covered by Miocene and Pliocene deposits. These deposits in turn are covered by sediments carried from the mountains by the numerous wadis. Coral reefs lie offshore along the coast. Where the largest wadis enter the sea their runoff has inhibited the growth of corals, so ports were located in such spots. Historically, the cities built on the Red Sea served as ports for trade between Egypt and lands to the south, such as **Punt**, or to the east toward Arabia and India. Ancient Egyptians probably relied on a port near modern Quseir. Greeks and Romans used three main ports, including Myos Hormos (modern Quseir), Philoteras (near modern Port Safaga), and Berenike (modern Berenice). Muslims crossing the Red Sea to Arabia for pilgrimages to Mecca left from Suez, Quseir al-Qadim (8 km north of modern Quseir), or Aidab, which is located close to the Egyptian–Sudanese border.

The most important Greco-Roman port was Berenike, founded by Ptolemy II in 275 BCE. This port was located farther to the south than the other cities so the ships could avoid sailing against the north winds of the northern Red Sea. During the reign of the Roman emperor Augustus, as many as a hundred ships set sail from here each year for distant places to supply the Roman demands for luxury goods, spices, and incense. This trade increase was the result of the Romans learning the secret of the monsoon winds, which permitted ships to sail directly across open seas to India with favorable winds each way once each year. The northeast monsoon winds blow from November to March, while the southwest monsoon blows from April to November. According to John Ball (1942), the trading vessels would leave Berenike in mid-summer. It took a month to reach Aden in Arabia (modern Yemen) and another month and half to reach India. By leaving India in December or early January, the ships enjoyed a northeast wind across the Indian Ocean and a south or southwest wind in the Red Sea itself. Arab seafarers followed these same routes in later times.

Recent archaeological excavations in the ruins of Berenike have revealed that the Ptolemaic town enjoyed another period of activity under the Romans in the first century BCE and first century CE, and then after a few centuries of decline was active again in the fifth and sixth centuries. Buildings were constructed from blocks of gypsum (a Miocene evaporite) or the reef coral heads that are abundant along the shore. Unfired bricks were made with salty sand. Among the products passing through this port were spices, gemstones, refined copper, and luxury items such as pearls, shell, red coral, and ivory. Although the harbor at Berenike was protected on the north by the Ras Banas peninsula, it lay at the mouth of a wadi draining from the Red Sea Mountains. This carried silt into the harbor, which may have led to the city's decline.

The trade items that were landed at Berenike were loaded onto pack animals for the 200 or 300-km overland trip to the Nile. Throughout most of pharaonic history, donkeys were the only animals available to supplement human porters. But the Ptolemies introduced the camel to Egypt. Camels were especially useful on the desert treks, since each animal could carry a greater load than a donkey, and they were able to go for several days without drinking and could browse on wadi shrubs for food. Camels can also drink water that is too salty for humans or other animals. Although Aswan appears to be the closest Nile town to Berenike, the caravan routes led northwest to either Apollinopolis Magna (modern Idfu) or Coptos (modern Qift). Perhaps Aswan was considered too close to the often-unstable Nubian frontier. The trip from Berenike to Coptos took about twelve days. Along the desert track, the Greeks, and particularly the Romans, built thick-walled stations at intervals of less than one day's march. These stations (called *hydreumata*) provided shelter for the trade goods, pack animals, traders, and the soldiers who came along to ward off attacks from the native Bedouins. Within each station was a well and a brick reservoir made waterproof with a cement lining.

At all periods, the track from Qift to Quseir was one of the busiest of the Eastern Desert routes, since it involved traffic to mines and quarries in the desert interior as well as to the Red Sea coast. This 175-km trip took about five days. Along the route there were at least eight stations, as well as a series of some 65 towers, each within sight of the one before and after it. These towers functioned to send signals between the coast and the Nile. The ancient caravan routes shown on Figure 14.2 followed wadis wherever possible; transits from one wadi to another were made where the **watershed** was the lowest, with gentle ascents for pack animals. The routes sometimes detoured from a direct line to take advantage of wells or to access mines.

Like all the other coastal cities, ancient Berenike and Quseir were completely dependent on towns in the Nile Valley to supply them with food; only fish could be obtained on the coast. Water supplies were absent along the coast, so towns relied on wells dug in wadis up to several kilometers inland. Ground water collected in the wells and was carried to the coast by pack animals.

Satellite view of Lake Nasser, Toshka lakes, and the Nile Valley
north to Qena (NASA)

The late-Miocene Nile Canyon near Cairo may have resembled the Grand Canyon in Arizona (FP)

Pediplain landscape along Lake Nasser

Granite boulders along the Nile near Aswan (RS)

Coarse-grained red granite from
Aswan; the coin is 2.3 cm (MK)

Holes prepared for wedges to split a
granite block

Dolerite pounder used to
shape granite

Hole for a dovetail cramp in a cracked
statue of Ramesses II

Graffiti on a block of easily carved Nubian sandstone

Hatshepsut's Mortuary Temple, with the contact between Esna Shale (thinner, horizontal strata below) and Theban Limestone (thick, massive layers above) visible above third terrace (FP)

The northern colossal quartzite statue of Amenhotep III, repaired during the Roman period (RS)

Unfinished quartzite statue—probably New Kingdom, possibly representing Amenhotep III— found in the quarry of Gebel el-Ahmar, Cairo

Salt efflorescence on an inscribed sandstone wall in Karnak Temple;
note spalling on the lower course

Spalled ceiling, cracked pillar, and flood debris in KV 10, tomb of
Amenemesses, in the Valley of the Kings (RH)

Inscribed travertine façade, barque chapel
of Tuthmosis IV, Open Air Museum, Karnak

Reconstructed Red Chapel of
Hatshepsut, Open Air Museum, Karnak

Inscribed blocks of the Red Chapel: red quartzite above,
gray granodiorite below

Human-powered shadoofs are still
used to irrigate small gardens (RS)

Pigeons living in dovecotes such as
this provide food and fertilizer (RS)

A donkey-powered waterwheel lifts water from
a well to irrigate fields (RS)

Satellite view of the northern Nile Valley, the Fayum,
and the Delta (NASA)

Fine-grained limestone, probably from a quarry at Tura

Limestone filled with fossil nummulites on the Giza Plateau

Spalling of limestone caused by salt subflorescence or hydration of clay layers

Basalt paving block (showing saw marks) in Khufu's Mortuary Temple,
east of the Great Pyramid, Giza

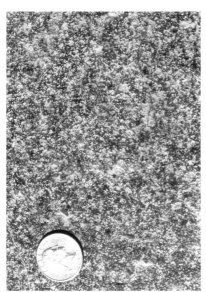

Fine-grained basalt;
the coin is 2.3 cm (MK)

Square stumps left after the removal of
limestone blocks northwest
of Khafre's Pyramid, Giza (MK)

The head and shoulders of the Sphinx show differential erosion as a result of different composition of various strata

The granite stela erected in front of the Sphinx by Tuthmosis IV in the Eighteenth Dynasty has been eroded by salt in rising ground water (CR)

Old Kingdom tombs at Giza show erosion similar in kind and degree to that on the Sphinx

Siwa Oasis viewed from the ruined village of Aghurmi (EG)

Wind-blown sand sculpted this limestone rock in the White Desert
near Farafra Oasis in the Western Desert (EG)

St. Anthony's Monastery backs against the escarpment of
the Southern Galala Plateau, Eastern Desert

Gezirat Fara'un (Pharaoh's Island) in the Gulf of Aqaba is composed
of granite, a small slice off the Sinai massif (MK)

Sunrise reveals the pinks, reds, and violets of the granite peaks
in southern Sinai (MK)

Satellite view of the Sinai Peninsula, Gulf of Suez (left),
Gulf of Aqaba (right), and northern end of the Red Sea (NASA)

The Red Sea

The warm, salty waters of this sea are normally very transparent, appearing blue not red, as they reflect the cloudless sky. Some have suggested that the name given to the sea comes from the colors of granites in the Red Sea Mountains reflected in the water; others say it refers to a bloom of red algae *(Trichodesmium erthraeum)* that sometimes tints the waters. Whatever the source of the name, the sea is a subject of intense interest to geologists, archaeologists, and travelers.

Because of its great length and southern extension, the waters are extremely warm, averaging about 22° C at the surface and up to 60°C (140°F) in the depths. No rivers regularly empty into the sea, and it receives a minimal amount of water from ephemeral wadis and occasional rainfall. Negligible water enters the Red Sea from the Mediterranean through the Suez Canal. The only inflow of water is from the Indian Ocean through the Strait of Mandeb. Since the rate of evaporation from the surface is high, the Red Sea's water is saltier than typical seawater.

Tides in the Red Sea average less than a meter, as a consequence of its narrow east–west dimension. There are several currents within the sea, which keep its waters well mixed. It is estimated that the water in the Red Sea is totally replaced every twenty years. Waters flowing inward from the Gulf of Aden are blown northward by prevailing winds, aiding mariners entering from the Indian Ocean. North of 19° N, though, the winds blow from the north. There are few natural harbors; most cities were located where runoff from wadis prevented the formation of coral reefs.

The Red Sea's Geological History

We learned in Chapter 3 that the formation of Red Sea began in the late Oligocene Epoch, around 25 to 30 million years ago. At that time, rifting began to detach a piece of the African Plate and form a small plate carrying the Arabian Peninsula. This plate first rotated counterclockwise (opening the Red Sea to a width of about 60 km); then the plate moved 100 km to the northeast, widening the Red Sea and creating the Gulf of Aqaba.

The Red Sea is very deep, considering that its maximum width is only 350 km. Along both of its coasts are shelves consisting of thinned continental crust topped by 4 or 5 km of sediments. The water depth over these shelves ranges up to 1,200 m deep. There is a narrow trough 2,000 m deep down the middle of the sea. South of 20° N, new oceanic crust is forming from upwelling magma within this axial trough. The age of this ocean crust and the absence of Miocene sediments in the trough suggest that sea floor spreading began here about 5 million years ago. The process of sea floor spreading is gradually being extended toward the northwest into the rest of the sea. The axial trough is punctuated by deeper pools that contain hot salty brine and sediments containing a fortune in oxides of iron, manganese, lead, gold, silver, copper, and zinc. Despite the allure of this rich resource, no method for recovery of these compounds has yet been devised.

The Gulf of Suez

The region encompassing the Gulf of Suez and the country to either side of it is intensely faulted. Major faults run northwest to southeast, paralleling the gulf, and these are crossed by many smaller faults that divide the crust into many separate blocks. For reasons that are not yet clear, the area now occupied by the Gulf of Suez was one of subsidence and deposition from at least the late Paleozoic Era onward, except for certain periods such as the Oligocene Epoch, when the region was high. The overall sequence of sedimentary rocks in this region is similar to that already described in the Nile Valley (see Fig. 2.2).

Fault blocks on both sides of the gulf displayed a somewhat independent behavior, however, in being uplifted and subsiding at various times, as evidenced by the variable thickness or even absence of certain sedimentary layers on one block compared to an adjacent one. Such independence of action might be explained by envisioning the fault blocks floating on the semi-liquid mantle. Erosion from a higher block onto a lower one would cause the lower

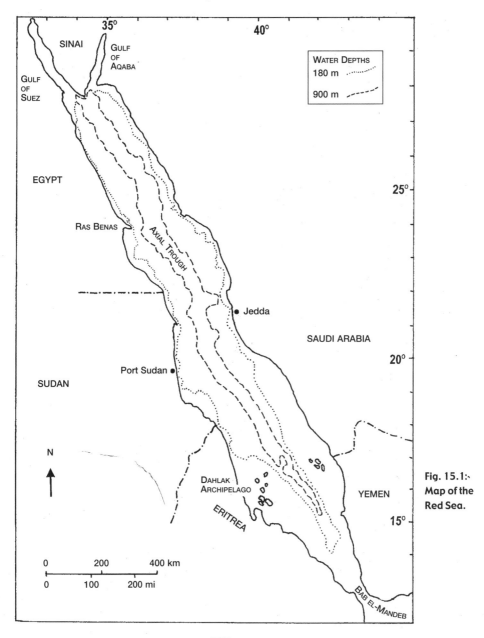

Fig. 15.1:
Map of the
Red Sea.

block to sink from the added weight, while the eroded block, although lowered at its surface by the loss of sediments would be uplifted by the process of **isostasy**.

The critical gulf-forming event of the late Oligocene–early Miocene may have been the onset of extensive tectonic forces that permitted the blocks—already defined by faults—to sink and form a basin. Blocks along the sides of the basin dropped less, and some were tilted and rotated. Lava erupted along many fault lines, forming sheets of basalt at about this same period. The Gulf of Suez was soon filled by water from the Mediterranean. Coarse sediments were carried from the Red Sea Mountains by runoff along the channels that survive today as dry wadis. These land-derived deposits were followed by ones of a marine character. These deposits now crop out on the coastal plains, showing that the Miocene gulf was somewhat wider than at present, with its coastline some 8 to 10 km inland. The thickness of the layers varies from point to point, indicating that the bottom of the gulf had an extremely varied topography, due to the varied elevations of the fault blocks. In some places coral reefs of this era occur 400 m above sea level. This elevation is more than can be explained by a former rise in sea level, and must show that the crustal blocks have themselves undergone uplift.

In the mid-Miocene, the Gulf of Suez was cut off from all but a trickle of water from the Mediterranean. This prevented the gulf from drying up altogether but caused the marine deposits to be succeeded by thick layers of evaporites, including halite (salt) and gypsum. In some places around the gulf, these deposits are several kilometers in thickness. These deposits are not contemporaneous with the evaporites of the Mediterranean Sea during its late Miocene desiccation, however, since water from the Mediterranean was needed to supply the gulf's shallow lakes. This layer of gypsum, along with coral-formed limestones, served as one of the major building materials of the coastal towns in ancient and medieval times.

By the Pliocene Epoch the Isthmus of Suez had risen and entirely cut the connection between the Mediterranean and the Red Sea. Meanwhile, the rotation of the Arabian block had created the Gulf of Aden and opened the Red Sea to the Indian Ocean. Pliocene deposits along the Red Sea and the Gulf of Suez include fossils reflecting this Indian Ocean connection. The Pliocene Gulf of Suez was a shallow lagoon, and its deposits were gravel and sand, not deep marine mud. The region must have been sinking, however, because the

Pliocene/Pleistocene deposits in the gulf are 1,500 m thick. The rate of sinking and of deposition must have been similar, thereby maintaining the shallow depth of the water for an extended period.

Today the gulf is less than 80 m deep throughout most of its length. The depth increases as one approaches the Strait of Gubal, and within the Red Sea the depth increases abruptly to over 1,000 m in the extension of the axial trough that reaches nearly to Ras Muhammad. We know that the sea level has fallen and risen many times in the past. During the Ice Ages of the Pleistocene, sea levels were as much as 100 m lower than today. During such times the floor of the Gulf of Suez would have been an exposed plain. Higher water levels are documented by gravel terraces (former beaches) at several points above sea level. On these terraces, archaeologists have found the remains of flint tools, which they have dated to several Paleolithic and Neolithic periods of human culture. There are also remains of coral reefs—formed during periods of higher water—along the coast.

The Gulf of Aqaba

In contrast to the shallow Gulf of Suez, the Gulf of Aqaba is very deep—averaging 1,250 m deep and reaching 1,700 m in its central trough. The difference in the depths of the two gulfs is explained by their mechanisms of formation. About 18 million years ago, a northeast-trending fault sliced through the northern Red Sea and created the Gulf of Aqaba. The Arabian block on the east side of the fault was pulled northward past the Sinai portion on the west for some 60 km during the Miocene, and another 45 km during the Pliocene/Pleistocene. There are no Miocene or Pliocene rocks in this gulf, indicating that its flooding did not occur until the Pleistocene. This sliding increased the width of the Red Sea, but had no effect on the width of the Gulf of Suez, which had been effectively isolated by this fault. This sliding motion continues at a rate of about 15 mm per year.

Seismic Activity in the Red Sea

Globally, most earthquakes occur along the boundaries of the major crustal plates. They are also associated with faults and areas undergoing rifting. In Africa, the highest frequency and greatest severity of quakes are observed along the Rift Valley system of East Africa. Egypt is considered rather quiet by world standards, but it has experienced a number of major quakes, and some areas of the country are more quake-prone than others. These areas include Aswan, the northern Red Sea, and northeastern Egypt from the Fayum north through the Delta. As recently as November 1992, a major earthquake caused extensive damage and 450 deaths in Cairo. Microearthquake activity is very common in the northern Red Sea, where the Gulf of Aqaba shear intersects the Red Sea's axial trough. This is also an area with a history of intermediate (magnitude 4.0 to 6.0 Richter) tremors. The earthquakes are almost certainly the consequence of the rifting still occurring in this region. Earthquakes have also been recorded in the rift valley that extends northeast from the Gulf of Aqaba, where two plates are grinding past one another—a situation similar to that occurring along the San Andreas Fault in California.

The Red Sea Coastal Plain and Islands

Between the high mountain ranges that border the Red Sea on both coasts and the sea itself is a coastal plain of varying width. The plain is rather flat except for old gravel terraces and outcroppings of basement or sedimentary rocks. These outcroppings represent the tops of displaced fault blocks separated from the main body of the mountain ranges that terminate in high escarpments along the coasts. For example, the Ash el-Milaha range northwest of Hurghada can be seen to be a gigantic fault block that has slipped and rotated, with its top now tilted toward the west. Within the block the usual vertical sequence of rocks are now lying side by side so that they reveal basement complex, Nubian sandstone, and Eocene limestones in parallel strips from east to west. This and other slipped and tilted fault blocks have been covered by deposits of the Miocene Epoch and later. The latest deposits are Pleistocene sediments carried over the plain by runoff from the numerous wadis draining from the mountains. The final component of the coastal plain and the islands was contributed by living corals.

Many islands lie along the Red Sea coasts. They were formed by a variety of geological processes. In the southern end of the sea, the island groups of Zubair and Zuqar were formed by volcanoes similar to those in Yemen, the Afar Triangle, and in the interior of Ethiopia along the rift valley. Gebel Tair lies on the edge of the axial trough. It is a still-active shield volcano about 4 km across.

Zabargad Island (also called St. John's Island) lies 80 km southeast of the peninsula of Ras Banas. Both of these areas have outcroppings of igneous and metamorphic rocks, but whereas the rocks at Ras Banas are exposed parts of

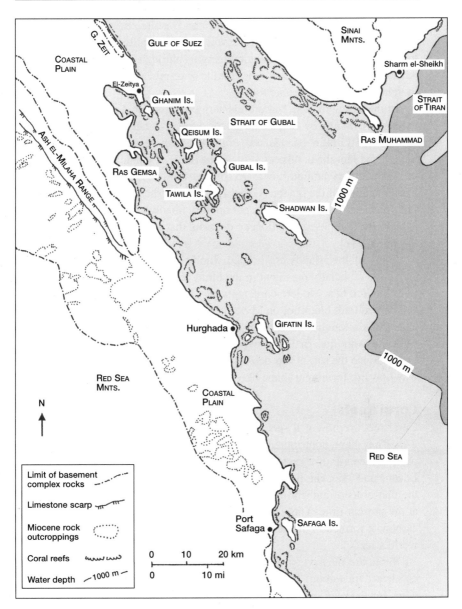

Fig. 15.2: The northwestern end of the Red Sea, with its many islands and coral reefs.

the ancient basement complex of the African Plate, recent studies suggest that some of Zabargad's rocks are relatively recently uplifted samples from the lower crust and the upper mantle emerging through rifts in the crust. This island was famous in ancient times for the semi-precious green peridots found in veins of the metamorphic rocks.

In the northern Red Sea are clusters of islands that appear either to be extensions of nearby headlands or the tops of other sunken fault blocks. Safaga and Gifatin Islands are examples with related headlands at Port Safaga and Hurghada respectively. In the Strait of Gubal there are numerous islands with a general northwest to southeast trend (Fig. 15.2). These may be a prolongation of the igneous Gebel Zeit ridge or the tops of other submerged fault blocks. Pink granite crops out on hilly Shadwan Island, while the other islands are composed of Miocene deposits that may cap older rocks below the water level.

Many of the islands are formed wholly or partially of coral limestone. These are especially prevalent in the Strait of Gubal, where the Gulf of Suez joins the Red Sea. Modern coral reefs may be growing atop the older rocks of submerged fault blocks or on Miocene reef platforms. Others may find a suitable shallow environment atop a salt dome. These salt domes formed as Miocene evaporites (halite or **gypsum**) rose through overlying younger sediments. Since the salt is less dense than the sediments, it gradually worked its way upward, forming a shape like an upside-down teardrop.

Coral Reefs

Coral reefs represent a special type of sedimentary environment (see Fig. 2.1). They have contributed to rock formation during many periods in the past; in the case of Egypt, two important periods of reef formation were the Eocene and Miocene Epochs. Many of the modern reefs may be growing on the platforms of ancient ones. Reefs are forming along most tropical coasts at the present time. These reefs produce more than half of the total calcium carbonate formed in the oceans today. The Red Sea contains 9 percent of the world's reefs.

Reefs are massive structures composed primarily of limestone. Living corals produce about 10 percent of this limestone, which forms the framework for the reef system. The framework provides support for many other limestone-secreting organisms and acts as a giant sieve to trap debris in the form of shells, skeletons, and spicules from the marine environment, as well as sediments washed from the nearby land. This debris is cemented to the reef by the action of limestone-secreting algae that grow in sheets over the reef.

The reef provides living space, food, and shelter for thousands of species of microbes, plankton, invertebrates, and vertebrates that form an incredibly complex ecosystem. In fact, reefs are among the most productive ecosystems in the world in terms of rate of formation of biomass. Like all ecosystems, the base of the food chain depends on the **primary producers:** the photosynthetic bacteria, algae, and plants. These provide food for the animal consumers. Interestingly, some animals, including many corals, have a symbiotic relationship with some type of primary producer. Coral polyps catch zooplankton with their stinging tentacles, but they also depend on high energy nutrients produced by symbiotic algae within their cells. It is the algae's need for sunlight that restricts the coral reef to shallow water—most often found along a continental coastline or volcanic island. Corals and algae can grow to a depth of about 80 m in clear water, but the optimum growth occurs above 30 m. Coral species without symbiotic algae can grow in the dark at great depths, but they grow very slowly. Since corals cannot grow above the sea level and are eroded by wave action when exposed at low tide, they tend to form a wide platform just below the surface of the water.

Reef-forming corals are colonial invertebrates. Each individual polyp within the colony extracts calcium carbonate from seawater (which is usually supersaturated in this compound) and secretes it as cups, tubes, or strands. Under optimal conditions, corals with a linear growth habit can grow as much as 10 cm per year. Compact corals may grow 0.5 cm annually. Corals are found only in seas where water temperatures never fall below 18° C. Warmer water, about 25° C is optimal for their growth, but they cannot survive excessively high temperatures either. The Red Sea averages around 20° C in winter and 21° to 26° C in summer. The sea's water is very clear, since there is little runoff from the wadis or pollution from ships, industries, or towns—at least not so far. Corals need clear, saline water, so they are missing at the mouths of major wadis because of sediment and occasional fresh water runoff.

Corals are not the only reef builders, although their frameworks form the basis of all modern reefs. A variety of algae also produce calcium carbonate. And single-celled *Foraminifera* are the third greatest producers of limestone in the reef community today after corals and algae. These organisms live in coral sands and contain algal symbionts that contribute to their great productivity. The *Nummulites gizehensis*, whose tests or shells are preserved in such great abundance in the Eocene limestones, were *Foraminifera*. Coral sands and other sediments at the base of the reef are an important and highly productive part of the ecosystem. In these sands much of the detritus from the reef is degraded by bacteria that recycle the nutrients back into the reef. The coral sands are produced by the constant erosion of the reef by waves, storms, boat propellers, coral-eating fishes, and boring sponges. These sands and sediments form the calcareous ooze from which much deep-water limestone is formed.

Coral reefs become the basis for an elaborate ecosystem. Hundred of kinds of invertebrates, brilliant fish, turtles, many other kinds of organisms thrive on the Red Sea reefs, finding both shelter and food there. There is a huge diversity of species in the Red Sea, and 20 percent of the species are found nowhere else. This is what makes the area so appealing to divers. But careless visitors, polluted waters, and ships' anchors can easily damage corals. The need for conservation and regulation to protect this precious environment is recognized by the Egyptian government, and gradually it is taking action. Several protectorates have been set up on the Sinai coast in the Gulf of Aqaba, while the area around Ras Muhammad is a National Park.

Ras Muhammad is a famous diving location. The beauty of its reef is legendary. Because it is located at the north end of the Red Sea axial trough, deep water (more than 1,000 m deep) is close to shore. Thus divers can see reef species as well as fish from the deep-water realm. The reefs' value to humans is not just for recreation. The productivity of reef systems benefits the surrounding water, which receives nutrients from it. The young of many deep-water fishes live on reefs; thus reefs are essential to a strong fishing industry.

Reefs have many natural enemies: storms, waves, climate change, and sea level rise or fall. Biological threats include disease and predators such as the crown-of-thorns starfish *(Acanthaster planci)*. A proliferation of these starfish can devastate a reef, and marine biologists cannot explain such outbreaks. Reefs do have amazing powers of regeneration from episodic damage. Unfortunately many of the threats to modern reefs, including those in the Red Sea, arise from human-made hazards that seem bound to continue and increase.

Coastal Occupations

In the last chapter, on the Eastern Desert, we learned that throughout history the cities on the Red Sea coast were ports from which expeditions set sail and at which trade vessels arrived from distant countries. These towns suffered from a lack of fresh water, the absence of agricultural land, and the long distance from the urban centers in the Nile Valley. Most of them were abandoned as their harbors became filled with sediments. Today, new towns are booming with local industries and tourists. A pipeline carries Nile water from Qena to Port Safaga. Desalinization plants that provide abundant fresh water for drinking and industry supplement this. The Red Sea has become both a destination and a busy highway.

Among the commercial ventures are oil exploration and extraction, phosphate mining, and fishing. Hurghada is the center of the booming Red Sea oil operations. Most wells are in the Gulf of Suez, along its coasts, or in the northwest region of the Red Sea proper. The geological conditions in the gulf's long history were especially conducive to the formation and trapping of oil. Refineries are located at Suez, and pipelines run from Suez to Cairo and Alexandria. Thanks to these oil reserves, and wells in the northern part of the Western Desert, Egypt is not only self-sufficient in petroleum but an exporter as well.

Other industries are centered on several Red Sea towns. Phosphates are mined at several locations near Safaga and Quseir and shipped from both ports. Ras Gemsa produces phosphates and sulfur. The phosphate mines tap the same Cretaceous stratum that is exploited in the Nile Valley near Idfu and farther west in the Kharga Oasis. The Red Sea phosphates are primarily exported to the Far East, while the mines in the Nile Valley supply the domestic needs. Wheat imported from the United States and Australia is unloaded at Port Safaga. Manganese from the Sinai mines at Umm Bogma is shipped from Abu Zenima.

In addition to the shipping to and from these and other Red Sea ports such as Elat and Aqaba, the sea is traversed annually by thousands of ships on their way to or from the Suez Canal, which has a capacity of 25,000 ships per year. Because of the narrow straits and shallow depths exacerbated by coral reefs, accidents have been inevitable. Although some shipwrecks have become diving attractions, reef damage has also occurred. Shipping is one of the main threats to the Red Sea reefs. Ships run aground, drop anchors, rub reefs with propellers, discharge ballast waters and sewage, and leak oil. Fortunately, the

175

largest oil tankers generally make the trip from the Indian Ocean to Europe by going around the Cape of Good Hope rather than through the Suez Canal, but oil wells also present a hazard, and drilling mud is toxic to corals.

The burgeoning tourist industry of the Red Sea has geological factors to thank for its success. The configuration of the mountains and sea provides the warm dry air, the warm clear water, and the usually calm sea. The weathering of the mountains has provided gorgeous scenery and wide, sandy beaches. At the Ain Sukhna resort, only 55 km south of Suez, the hot springs are an attraction. And offshore, the coral reefs constitute a marine wonderland—a stark contrast with the sterile deserts. These same tourists who are attracted to the reefs also threaten them. Unsupervised divers and fish collectors remove reef organisms. Desalinization plants return hot salty water to the sea; this briny discharge can kill corals at distances up to 2 km away. Sewage effluent from hotels or ships provides extra nutrients to the reef ecosystem, which encourages the growth of seaweed that inhibits and gradually replaces corals. Without the corals, the community declines. It is to be hoped that Egypt, for which tourism is a significant source of revenue, will not allow popularity to kill the goose that lays the golden eggs.

The Isthmus of Suez and the Suez Canal

Geologically, the Suez Canal can be considered an artificial continuation of the Gulf of Suez. The gulf tapers out in the region of the Bitter Lakes; north of these lakes was a shallow trench in which the canal was dug. In fact, in the Miocene Epoch there was a continuous natural seaway connecting the Red Sea with the Mediterranean. A rise of sea level of only 20 m would again flood this trench.

As we have seen throughout this book, the physical environment had important consequences on human activities. The Isthmus of Suez—although inhospitable during historical times as a place for human habitation—was either a crucial bridge or an annoying barrier to the movement of migrants, armies, and traders. The construction of the Suez Canal converted a formidable land barrier to shipping into a vital maritime connection.

The idea of building a canal to link the Mediterranean and the Red Sea is very old. It may trace its inspiration to the canal the pharaoh Necho dug in about 600 BCE through the Wadi Tumilat to link the Nile and the Red Sea. Certainly during the fifteenth and sixteenth centuries CE Venetian traders would have benefited from a direct water route between the Mediterranean and the Red Sea to help them compete against the Dutch and English merchants who were using the route to India around the Cape of Good Hope discovered by the Portuguese. The longer water route was only adopted to avoid the tolls of Muslim potentates who controlled all the overland routes.

The north end of the Red Sea was characterized by "shallow water, shifting shoals, treacherous currents, consistently unfavorable head winds" (Holladay 1982: 33). So long as ships were rowed, this sea could be navigated. Once freighters were primarily sailing ships, however, ports farther south were sought even if this entailed a long overland caravan journey to reach the

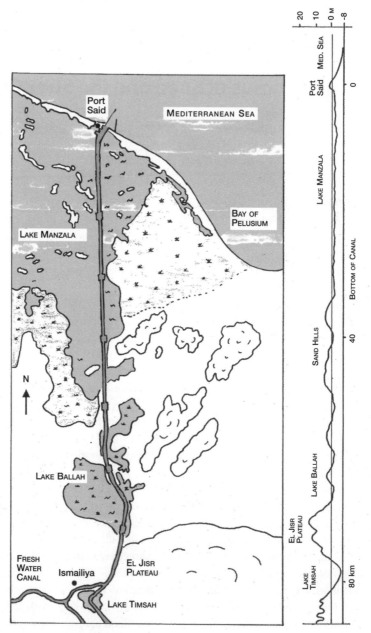

Fig. 16.1: Map of the Suez Canal as built in 1869 (northern half, left; southern half, right). A longitudinal section through the canal is to the right of each plan with sea level at 0 m and the bottom of the original canal at 8 m below sea level.

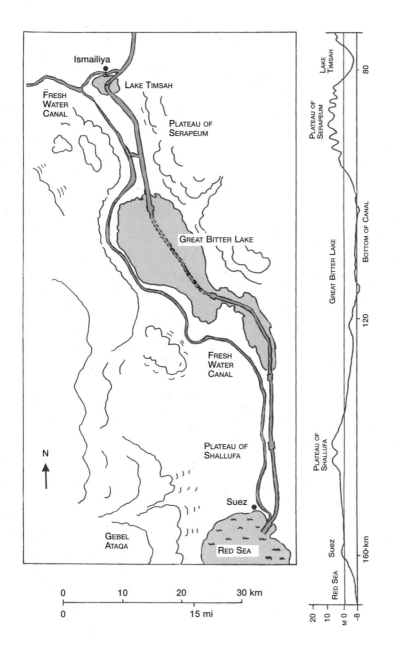

Nile and the Mediterranean. The development of steamships in the mid-nineteenth century eliminated the limitations imposed by the headwinds in the Red Sea; it therefore became an increasingly attractive route to the Mediterranean. But passengers and goods unloaded at the town of Suez still faced an arduous 130–kilometer desert journey to reach Cairo, where they could continue their journey down the Nile to Alexandria and board ships for Europe. Once again, nineteenth-century visionaries began to promote a sea-to-sea canal.

Napoleon's temporary conquest of Egypt in 1798 provided a real impetus for a continuous sea route, since he believed that such a canal would contribute to his goals of interfering with Britain's eastern imperial stakes, as well as assisting French merchants. He therefore sent his engineers to survey the isthmus. Unfortunately, they concluded incorrectly that the Red Sea level was 10 m higher than the Mediterranean. This error affected later estimates of a canal's feasibility and led some advocates to recommend following the ancient route from the Mediterranean up the Nile and through a newly-excavated canal in the Wadi Tumilat to the Red Sea.

In fact, it was to be another Frenchman, Ferdinand de Lesseps, who was to take up the challenge of building the lockless, sea level canal. The fascinating story of the Suez Canal's construction between 1859 and 1869, with the involved economic maneuvering and political intrigues among European powers, the Ottoman Porte, and Egypt's viceroys has been ably told by several authors (see Marlowe, 1964, or Kinross, 1969). When the canal opened, its effect was to reduce the distance between a port in the English Channel and the Far East by 8,000 km, or the equivalent of about 36 sailing days.

What sort of geological challenges did the canal builders face? The Isthmus of Suez between the Mediterranean and the Red Sea consisted of a series of depressions and ridges. The depressions were dry, although they lay partly below sea level. The construction involved cutting through the ridges and allowing the depressions to fill with seawater. A channel also had to be dug through some of the depressions to obtain sufficient depth. Along the Mediterranean coast and separated from the sea by a narrow sand bar was Lake Manzala, but it was too shallow to be navigated by large ships. Thus a channel had to be dredged through the lakebed.

Most of the ridges were composed of sand and alluvium that were easy to excavate, but at the Shallufa Ridge, 52,000 cubic meters of rock had to be removed by blasting. A smaller ridge of rock was also encountered at the

Serapeum. At the Jisr Ridge, too, difficulties occurred, since the sand dunes were nearly 20 m high, and the sand tended to slide back into any excavation. Eventually, enough sand was removed by manual labor to create a shallow trench; when this was flooded, floating mechanical dredgers—many invented specifically for this project—were able to complete the work. In all, 75 million cubic meters of material were excavated over the 171 km length of the canal from 1860 to 1869. This represents a volume equal to thirty Great Pyramids.

The dredged materials were used to construct the canal banks, to create additional land at the ends of the canal, or to produce concrete, depending on the location and the nature of the excavated material. Beneath the sand and lagoon mud, the builders discovered a layer of clay. When used for canal banks it dried into a hard impervious surface that served as an excellent roadway. The raised canal banks also helped prevent wind-blown sand from depositing in the canal.

Two other construction projects were necessary before the canal itself could be started: a Mediterranean port and a fresh water canal. The Mediterranean coastline presented no natural harbor. Indeed, the sea was shallow for a considerable distance from shore, the winds were northerly, and the currents were treacherous. Therefore, it was necessary to dredge a channel several miles in length out from the shore and to protect it with a breakwater extending over 2 km into the sea. On shore, port facilities were built at a new site called Port Said. This would eventually serve as the northern terminus of the canal, but during construction it provided a place to unload the construction materials. Stone for the breakwater was lacking in the immediate area, so limestone blocks were transported by ship from Mex, west of Alexandria. When this transport proved too costly, gigantic blocks of concrete were cast on location; nearly 30,000 22-ton blocks of concrete were made. After only a few decades these synthetic blocks began to disintegrate, however, and they were replaced with limestone blocks from quarries at Gebel Ataqa. This mountain is located west of Suez, and quarried blocks could be shipped north to Port Said on the canal.

At Suez, the historical terminus of eastern trade, there was only limited fresh water, and throughout the route of the canal there were essentially no natural supplies of fresh water. Therefore a canal had to be built to bring water from the Nile to supply the thousands of workers building the canal. This freshwater canal (called the Sweet Water Canal then and the Ismailiya Canal now) followed the old route through the Wadi Tumilat to Lake Timsah, then turned south and paralleled the ship canal as far as Suez. The completion of this canal not only transformed the Suez area but also permitted the cultivation of nearly 110 sq km in the Wadi Tumilat. Unfortunately, leakage of water from the canal into the surrounding soil soon flushed salts to the surface and degraded its agricultural potential. During construction of the maritime canal, the freshwater canal was also used to transport men and materials from Cairo toward the southern half of the project. Fresh water was conveyed by pipeline from the Lake Timsah junction northward to Port Said.

Like most canals, the Suez Canal had a trapezoidal cross section; it is wider at the water surface than at the bottom. The walls of the canal had a very shallow slope to prevent cave-ins. The narrower sections of the canal were those where excavations had to pass through the former ridges, a distance only 20 percent of the total, since it would have been too expensive to cut the channel wider in these areas. From the moment the canal was opened in 1869, however, maintenance and upgrades were undertaken. The improvements included widening and deepening the channels, digging more sidings, and adding stone facings to the canal banks. It also proved necessary to dredge the channels at Port Said yearly and to lengthen the breakwater.

From 1874 to 1951, more than 158 million cubic meters of additional material were removed by dredging and more than 30 million by excavation. After the Second World War a major renovation was undertaken to deepen the channels to accommodate larger ships, to improve the harbors, and to cut a by-pass canal one-third of the distance from Port Said to Suez. This by-pass created a second point (along with the Bitter Lakes) where convoys of ships could pass one another. Before the improvements, ships could not overtake or pass each other. When ships moving in opposite directions were about to pass, one of them had to wait in a siding—a wider area along the channel provided for this purpose. Ships were not allowed to transit at night.

The post-war improvements involved excavating or dredging an additional 40 million cubic meters of material. The canal was blocked by sunken ships during the 1967 war between Egypt and Israel. Egypt regained the possession of the canal after its 1973 attack on the Israel occupiers, and the canal reopened in 1975. Major improvements were also undertaken in the 1970s to permit even larger ships to pass through it. This work involved the excavation of an additional 230 million cubic meters of sand and rock. The absence of locks, a limiting factor on the size of transiting ships in most canals, has made

. the task of upgrading the Suez Canal easier. The most recent work on the canal has involved the construction of a highway tunnel under the canal and a bridge over it at el-Qantara. Another method for permitting larger ships to use the canal has been the construction of a pipeline from Suez to the Mediterranean. A fully loaded oil tanker—too deep to pass through the canal—can unload its oil at Suez, transit the canal unloaded, and pick up the oil at the Mediterranean port.

The dimensions of the canal as built in 1869 and as improved are given in Table 16.1. The Suez Canal Authority plans to deepen the canal to 22 m by 2010, which would permit 90 percent of the world's largest oil tankers to use the canal. In 2000–01, the canal's revenues were $2 billion, and 7 percent of the sea-transported world trade passed through it.

Table 16.1: Dimensions of the Suez Canal, in meters

	1869	1954	2010
Width at surface	60–100	150	320
Width at bottom	22	60	100 (est.)
Minimum depth	8	10 (14 in main channel)	22

The Sinai Peninsula

The Sinai Peninsula contains 61,000 sq km or about 6 percent of the Egyptian total land mass. It has a triangular shape with a base 200 km wide along the Mediterranean coast and an apex thrusting into the Red Sea 380 km to the south. Sinai shares many of the geological features of the Eastern Desert of Egypt as a consequence of having been part of a single landmass up to the mid-Tertiary Period. The peninsula was separated from the African continent to the west and the Arabian Peninsula to the east by **faults** associated with a system of **rift valleys**. Tectonic uplifting at the southern end then tilted the entire Sinai block toward the north.

The southern quarter of the peninsula is a **massif** composed of the ancient basement complex. About 25 percent of the rocks are metamorphic types, while the rest represent gigantic igneous **plutons** that were intruded into older rocks. More recent sedimentary rocks that originally covered the igneous rocks were removed by erosion—a process that produced the landscape of jagged peaks we see today. Several of these peaks rise higher than any in the Eastern Desert, with Gebel Zebir (of which Gebel Katherina is one peak) being the tallest at 2,642 m. While the granites form either rounded domes (pink granites) or jagged crests (red granites), some **dikes** of intruded rock are more resistant to weathering and produce sharp vertical ridges.

The southern block of mountains has been affected by tectonic movements of many periods. The rocks are intersected by dozens of faults, some of which run roughly north and south parallel to the two coasts, while others run from east to west. These faults served as conduits for mineral-rich quartz that formed veins containing copper, manganese, and iron ores; they also directed the courses of wadis.

The many wadis that dissect the mountains have carved steep gorges and

184

The Sinai Peninsula

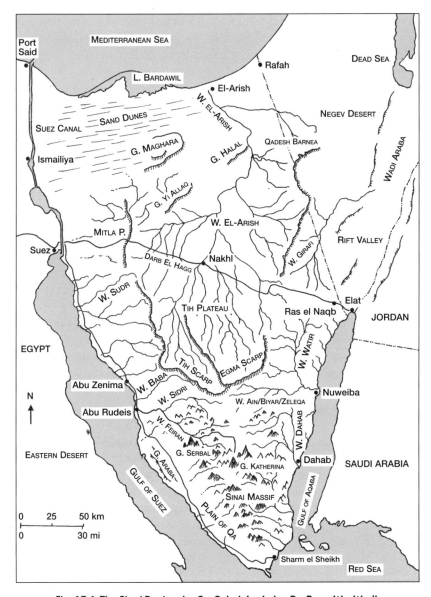

Fig. 17.1: The Sinai Peninsula. G = Gebel, L = Lake, P = Pass, W = Wadi.

185

ravines. This region receives more rainfall than any other part of the peninsula except for the Mediterranean coastline—an average of 50 to 100 mm a year. The rain is mostly in the form of local thunderstorms that produce raging torrents in some of the wadis. Rain soaking into the sediments on wadi floors supplies some small oases, as does snow melt from some of the peaks. Winter temperatures in the mountains often drop below freezing at night; this is the only place in Egypt where rock weathering resulting from freezing and thawing regularly occurs.

The wadis drain either to the Gulf of Suez on the west or to the Gulf of Aqaba on the east. Wadis draining westward from the southern massif have deposited sediments that form the wide coastal Qa Plain between the mountains and the Gulf of Suez. When water flows from the wadis, it usually soaks into the plain and does not reach the coastline. The Qa Plain, stretching north and south for about 130 km, has been considered for agriculture since it has a large proportion of all the soil suitable for cultivation on the peninsula. Unfortunately, the water supply from the wadis is not reliable, and rainfall over the plain averages only 10 to 20 mm a year. No such plain occurs on the eastern side of the peninsula, where the mountains plunge abruptly into the water of the Gulf of Aqaba and the wadis are especially steep, but Wadis Watir, Dahab, and Kid have formed small deltas at their mouths. Towns are located on these small flat plains and derive some water from wells dug into the wadi sediments. This water source is not sufficient, however, for the influx of tourists who now flock to the hotels erected here.

North of the Sinai Mountains in the center of the peninsula is the Tih Plateau occupying about 20 percent of the peninsula. Its surface is composed of nearly horizontal Cretaceous and Eocene limestones, whose strata show a slight dip toward the north. The Tih Plateau has an elevation of nearly 1,000 m on its southern edge, with Gebel Egma, an uneroded remnant, rising from its center to 1,626 m above sea level. The Eocene limestone of Gebel Egma contains excellent seams of flint that was much sought after by early Sinai inhabitants. The eastern and western escarpments of the Tih Plateau are defined by faults and represent the edges of the rift valleys occupied by the Gulfs of Suez and Aqaba. Just as on the western side of the Gulf of Suez, fault blocks on the east side have dropped or tilted to form a series of jagged ridges stepping down toward the gulf. Gebel Araba, formed from a fault block, parallels the northwest–southeast trend of the Gulf of Suez in the same way that Gebel Zeit and Ash el-Milaha do on the west side of the gulf.

The drainage channels of the vast Tih Plateau are wide and shallow, since it has such a gradual slope toward the Mediterranean. Many of the wadis beginning near its southern end coalesce into the north-flowing Wadi el-Arish, which has a drainage basin of nearly 20,000 sq km and a length of 310 km. The tributaries of this wadi form a huge ramifying system over the surface of the plateau. Greater rainfall during the Pleistocene Epoch, or even earlier, was responsible for transporting the vast quantity of sediment that lies in the beds of these dry channels. This is a second area on the Peninsula where potentially fertile soil occurs, but where the absence of present-day water sources prevents its exploitation. Since the potential for evaporation of moisture from the soil always exceeds the potential rainfall, there is no ground water reservoir to support perennial plants.

On the southern side of the central plateau are a steep south-facing escarpment and a series of valleys in which the Nubian sandstone—normally expected between the ancient basement complex and the more recent limestone strata—is clearly exposed. The sandstone has been deeply eroded by the west-flowing Wadis Baba, Sidri, and Feiran as well as by the east-flowing Wadis Watir, Ain, Biyar, and Zeleqa. Many mines both ancient and modern are located in this colorful 'sandstone belt' that runs from west to east between the Tih escarpment to the north and the Sinai mountains to the south. In the mining district of Umm Bogma, limestones and sandstones dating to the Paleozoic Era (Carboniferous Period) are exposed beneath the Nubian sandstone. The commercial ores of iron and manganese oxides are found at the base of the Carboniferous limestone stratum. Two important minerals that occur within joints within the Carboniferous sandstone are turquoise and malachite, which the ancient Egyptians prized for jewelry. The only place that sandstone of this great age is known in the rest of Egypt is in the Wadi Araba of the Eastern Desert, but no minerals have been discovered there.

North of the central plateau is an area crossed by faults and folds, both of which run in a southwest to northeast direction. The folds were formed when the African crustal plate began to collide with the Eurasian plate. The tops of some folds appear as high limestone hills such as Gebels Maghara, Yi Allaq, and Halal. They are surrounded by erosional debris and a rolling plain covered with sand. Closer to the coast, these folds may continue, but their presence is hidden by the coastal dunes that can rise as high as 80 to 100 m. The erosion of the tops of the hills has revealed older sedimentary rocks that would normally lie deep underground. In Gebel Maghara, rocks of Mesozoic (Jurassic) age have been revealed that contain commercially usable quantities of coal.

A wide plain flanks the Mediterranean coastline from the Suez Canal to Rafah. This plain has sand dunes running parallel to the coast and extending inland for 20 to 50 km. These dunes collect the rainfall from the winter rains, which average 50 to 100 mm per year. The water accumulated under the dunes supports some agriculture and supplies shallow wells. Lake Bardawil is a lagoon separated from the Mediterranean by a sand bar (in some places one kilometer wide) that was formed from Nile sediments debouched by the Pelusiac and other eastern branches and redistributed by eastward-flowing ocean currents. Artificial openings through the sand bar permit seawater to flow into the lagoon, where fishing is a major industry.

Human Occupation

The Sinai Peninsula's geography unites it to both Egypt's Eastern Desert and the Negev Desert of Israel. As the only land bridge between the African and Asian continents, it has been the route of migrants, traders, and military expeditions throughout human history. It has firm cultural links to both Egypt and the Near East. Today, much of the peninsula is an inhospitable place, but it had denser occupation during prehistoric times when the conditions were likely more equitable. At all times, human occupation was constrained by the geological features of the area: the topography, the kinds of rocks and minerals, and the sources of water.

Archaeological surveys along the Mediterranean coast reveal that this area has been populated since at least the fourth millennium BCE. This occupation was possible because ground water is trapped by the coastal dunes and can be accessed by wells. Winter rainfall can also be trapped and stored in cisterns excavated in the bedrock. In the early twentieth century, geologist William Hume observed several cisterns near Wadi Umm Khisheib and noted that they were excavated into the easily-worked Cretaceous limestone. Blocks from the excavation had been used to dam a nearby wadi and direct the runoff from episodic rainstorms into the cisterns. He estimated their capacities at from nearly 200 to over 1,500 cubic meters.

Remains in the coastal region from the Predynastic era reveal cultural influences from the Levant making their way toward Egypt, as well as expansion by ancient Egyptians toward the east. This evidence dwindles during the Second Dynasty, and it may be that Old Kingdom contacts were conducted by sea rather than land. The absence of Middle Kingdom remains implies the same thing, but beginning in the New Kingdom, when Egypt became an imperial power, the coastal road became a royal highway. This route left the Eastern Delta at Sile (modern el-Qantara) and proceeded toward Rafah; it was called the Way of Horus or the King's Highway. Forts were erected to control and assist the movement of armies and traders. After the conquest of Egypt by Rome in 30 BCE, the road across Sinai was improved with way stations to provide shelter and changes of horses.

Evidence of prehistoric occupation has been found at several non-coastal sites where permanent springs or ponds existed. These sites are located in Wadi Feiran in the southern massif, Gebel el-Maghara in the northern plain, and Qadesh Barnea in the upper reaches of the Wadi el-Arish basin. Archaeologists found a variety of flint tools or flakes from tool making that show an affinity to ones in Israel. Charcoal from several hearths was dated to between 34,000 and 30,000 years before the present. Remains of bones from some locations suggest a hunter–gatherer mode of subsistence.

Beno Rothenberg found evidence for two later periods of very intense occupation that he has termed the Elatian and the Timnian. These periods have been dated 4500 to 3800 BCE and 3800 to 2600 BCE respectively. It is probable that Sinai received more rainfall at that period than it does today.

Numerous Elatian settlement remains were found on both the eastern and western edges of Tih Plateau, where water was available in springs and wadis. The arid central desert was avoided as a site for settlements, but may have been used for pasture. The Elatians were semi-nomadic hunters and shepherds who occasionally cultivated wadi floors. Their chief mineralogical interests were flint; turquoise and malachite for ornaments, pigments, and eye paint; and native copper. Small nuggets of pure copper could be dug from a rock face and hammered into objects. This was probably the source of beads that have been found in Egyptian graves of this period. Along the southern edge of the Tih Plateau at Gebel Egma, flint deposits were mined and tools produced. Similarities between Elatian sites and those in Arabia indicate that the Elatians probably originated to the northeast of Sinai.

Timnian settlement sites occur over a wide area of eastern, southern, and western Sinai. These people were semi-nomadic herders who also raised grain. The Timnians were also the first to smelt copper from ore they mined, and many settlements were discovered in the regions of richest ore deposits. They collected and worked flint in a style showing Egyptian and Nubian influences. Some of their household pottery was imported from Egypt. Finds of copper and turquoise in Egypt in this period indicate an active trade between the two areas. Whereas stone huts of the Elatian period were round, the latest Timnian houses were more rectilinear, like the Egyptians'. Rothenberg sees all of this as evidence for a "homogeneous cultural zone" encompassing Lower Egypt and Sinai, whose inhabitants were conquered by Upper Egyptian forces during the unification of the country. This conquest, perhaps combined with environmental changes, resulted in the nearly complete disappearance of the Timnian Culture in Sinai.

The Timnian inhabitants were succeeded by waves of nomadic herders moving from the northeast, who occupied many previously settled sites for a short period of time. Rothenberg, who has investigated the archaeological evidence for this migration, equates it with the Biblical Amorites, who caused the collapse of the Sumerian kingdom in Mesopotamia; overran the lands of modern Syria, Palestine, and Israel; and finally moved into Lower Egypt during the disorder of the First Intermediate Period.

Beginning in the Third Dynasty of the Old Kingdom, the Egyptian kings sent expeditions to Sinai to obtain copper, turquoise, and other minerals. Late Timnian copper smelting sites show evidence of increased contact with and influence by Egyptians. The Egyptians showed little interest in colonization to replace the Timnian settlers, however. Archaeological investigations of Egyptian sites reveal only the shelters and work floors needed for miners rather than whole settlements with houses, cemeteries, and cult centers as in the Timnian period. The concentrations of desirable minerals were much lower than modern methods require for extraction, but the ancients patiently picked out pieces of gems and ore by hand. Turquoise mines were located in the southwest part of the peninsula in the Wadi Maghara (this is not the same as Gebel el-Maghara on the northern plain) and at Serabit el-Khadem. Copper ore was mined at Bir Nasib. Mining and smelting of copper ores continued in the Middle Kingdom and reached a peak during the New Kingdom. It has been estimated that 5,000 tons of metallic copper were removed from mines near Bir Nasib alone. Egyptian exploitation seems to have halted after Ramesses III.

The tradition of Christian monasticism that originated in Upper Egypt and the Eastern Desert spread to Sinai. Hermits sought out remote locations in the southern massif to build huts or live in caves. The Monastery of St. Catherine was built in the sixth century CE at the base of Gebel Musa and has operated ever since. Christian pilgrims were attracted to this mountain on which tradition says that Moses received the Ten Commandments.

After the Muslim conquest of Egypt in 641 CE, a route was developed across Sinai to accommodate the pilgrims going from Egypt to Mecca. They followed the Darb el-Hagg southeast from Suez to Elat and then southward into Saudi Arabia. This route across the peninsula followed that of even earlier Nabataean traders, who were active in Hellenistic times. Cisterns in this region may date from the Nabataean period, since they were excellent hydrological engineers. The pilgrims' path climbed the steep cliffs along the western edge of Tih Plateau, reaching it via the Mitla Pass. South of this pass is Wadi Sudr, with many springs and wells. The road led eastward to Nakhl, with bountiful wells in the valley of Wadi el-Arish. On the east edge of plateau the road crossed the steep cliffs of Ras el-Naqb, then descended to Elat. Fortresses were built at several points to protect the pilgrims. A modern highway now follows this route.

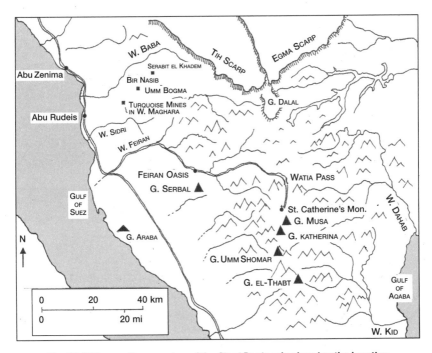

Fig. 17.2: The southern region of the Sinai Peninsula showing the location of mines and a route to St. Catherine's Monastery. G = Gebel, W = Wadi.

191

Modern travelers will find many attractions in Sinai, the majority of which are the consequences of geological processes and features. These include the dramatic scenery of the Sinai Mountains and the coral reefs along the coasts of both the Gulf of Suez and the Gulf of Aqaba that were described in some detail in the chapter on the Red Sea. Ras Muhammad, one of the premier diving locations in the world, is a peninsula formed from Miocene Epoch reefs. Another unusual spectacle is the 'Colored Canyon,' northwest of Nuweiba via the Wadi Watir. This narrow canyon was carved into colorful Cretaceous sandstone by running water during a wetter period. The stratum corresponds to the Nubian sandstone of the Egyptian mainland; it owes its colors to the weathering of the iron and manganese oxides naturally present in the sandstone.

A traveler heading from Suez to St. Catherine's passes an amazing array of rocks and landforms. The scenery is somewhat tedious at first as the road crosses the coastal plain of sediments washed from the distant hills. Small hillocks of Miocene deposits, mostly reefs and coarse-grained sedimentary rocks, offer some relief. On a clear day, one can see the flat tops of the two immense Galala plateaus across the shimmering Gulf of Suez. Then the road turns inland to skirt the Eocene limestone outcropping forming the Gebel Hammam Faraun (Mountain of the Pharaoh's Bath), with hot springs issuing from its base. This 500-meter peak is a notable landmark on the Sinai coast when viewed from the highway on the west side of the Gulf.

As the road leads farther inland, more hills of limestone and colored sandstone appear; some of these have tilted strata, indicating that they are the tops of tilted crustal blocks marking the east side of the Gulf of Suez rift valley. As the road leads into the Wadi Feiran, ancient rocks of the basement complex appear on either side. In its western end, the wadi cuts through **gneiss**, a metamorphic rock formed from even older rocks. Some authors give an age of 1,100 million years for this rock because of its similarity to rocks in the Eastern Desert. Laboratory tests have shown that it underwent metamorphosis around 630 million years old, while some granodiorites to the east were formed 780 million years ago. Together these rocks are among the oldest in the peninsula.

We have already learned how geologists use the 'law of superposition' to date rocks relative to one another. This works well with sedimentary rocks in which younger layers are deposited sequentially on top of older layers. Intrusive igneous rocks turn this rule upside down, since they are formed from magma rising into older rocks. Intrusions may take the form of huge **plutons** of material that remain fairly self-contained as they rise into the older rocks; or the magma may rise through fractures in the older rock to form vertical **dikes** or intrude itself horizontally between strata to form a sill. In the field, a geologist attempts to determine the relative age of igneous rocks by seeing which rock has intruded itself into another one. Because of the great age of the basement complex in Sinai and the many episodes of igneous rock formation, this situation is extremely complicated. Fortunately, age relationship based on field observations can be confirmed or corrected by using laboratory methods to determine absolute dates in some cases (see Chapter 1).

The ancient gneiss of the Wadi Feiran is crisscrossed by hundreds of dikes, many of which appear in nearly parallel lines of similar color and texture. These are referred to as dike sets or swarms and indicate that a single body of magma flowed into a set of pre-existing fractures with a similar orientation. These fractures were probably the result of compression, tension, or other tectonic forces. The rock in the dikes has been tested and dated to at least three different periods. The oldest dikes, which are primarily andesite, date to 591 million years before the present. These ancient dikes are cut by the next youngest set, which are primarily basalts of various types. Both of these ancient sets trend northeast to southwest. A final set, much less numerous than the previous two, is primarily composed of rhyolite; these dikes trend northwest to southeast, parallel to the Red Sea rift, and may date from the Miocene Epoch. We see that some of the dike rocks appear to be denser and more resistant to erosion than the rocks they intrude. They seem to shatter easily and weather into sharp-edged fragments instead of into rounded boulders. They often form ridges along the tops of a series of peaks and can be traced for many miles.

South of the row of grayish, gneiss peaks that parallel the Wadi Feiran rises the enormous pink granite bulk of Gebel Serbal. Its ridge extends for nearly 5 km, and in a manner characteristic of this kind of rock has weathered into 10 separate knife-edged peaks. This mountain belongs to a tectonic phase that has been dated to between 600 and 570 million years ago. Gebel Serbal overlooks the Feiran Oasis, the most extensive oasis on the Peninsula. Water to support the palm trees and other vegetation comes from a spring that flows for a short way on the surface and from wells that penetrate to the water table below the wadi.

The flat bottoms of the wadis are a surprise to many travelers who are accustomed to the V-shaped river valleys of wetter climates. The wadis are narrow, steep, and nearly impassable in their upper reaches, but in the lower stretches where the grades are gentler, the valley is wider, and coarse sediments completely fill the valley floor. Water flows in a wadi only occasionally, following a cloudburst over its catchment area; since very little runoff is absorbed by the hard bare rocks of the mountains, most water drains into the wadis. These brief floods move the sediments along and redistribute them across the floor. The large size of some of the boulders testifies to the power of these infrequent but potentially dangerous floods. Before the flowing water can carry the debris out of the wadi's mouth, however, the water sinks into the deep sediment layer of the wadi floor, where it forms a long-lasting reservoir that supports perennial vegetation. Tributary or side wadis usually drop their sediments and form alluvial fans where they enter a main wadi.

Beginning at the Feiran Oasis, a different kind of rock formation appears in places along the cliffs. This yellowish, obviously sedimentary deposit in some places reaches a height of 18 to 30 m and continues for several kilometers to the east. It is clearly different from the boulder-filled alluvial fans that appear at the mouth of a tributary wadi. Many geologists think these deposits were laid down in a lake that formed when something (possibly a landslide) blocked the valley of the wadi just west of the Feiran Oasis. The sediments have been dated to 60,000 years old (at the bottom) and 12,000 years old (at the top), a period corresponding to one of the Pleistocene **Pluvials**. The layered appearance of the deposits resulted from periodic floods that brought loads of sand and silts into the lake; the layers of clay and marl indicate years when the lake was deep and the water sat for longer periods, allowing these fine particles to settle. Eventually, the material damming the wadi eroded away and the floods began to erode the lake deposits. A meandering channel was cut down through the deposits, which preserved patches of the lake sediments along the sides of the wadis. Erosion has carved this remaining soft material into fantastic spires and towers that resemble gigantic sand castles. Bedouins have also dug caves into the soft stone, while in other places they have built on top of the deposits at a safe height above the present floods.

As the road nears St. Catherine's, it passes through a high wall of serrated, red granite peaks. This 'wall' is one of a series of parallel granite dikes running northward from the central mountain massif and now exposed by erosion. These huge dikes are especially clear on satellite photographs. Other prominent landscape features around St. Catherine's are portions of a geological phenomenon called a ring dike. As the name implies, this is a dike that forms a circle, or in this case an ellipse 27 km long from east to west, and 20 km wide from north to south. St. Catherine's Monastery lies roughly at the center of the ellipse. The dike is 2 km thick at its widest point, but only a few tens of meters thick in some places; it disappears completely, supplanted by more recent material, along parts of its course. Ring dikes are formed when an uplifted dome of rock or the top of a volcano collapses, creating a series of fractures arranged in a rough circle, through which magma can rise.

Geologists studying this ring dike have concluded that a cylindrical structure formed deep underground about 597 million years ago and that it was roofed over with similar igneous material, which cooled at a depth of several kilometers under the ground. Later uplift of the area, coupled with erosion, exposed the dike to view at its present elevation. Portions of the roof of this structure survive as the dark tops of several peaks, including Gebels Musa, Katherina, and Abbas Pasha. The lower bulk of these and the entirety of other peaks within the ring complex are derived from a huge pluton of red granite that was intruded into and under the ring dike rock about 580 million years ago: another example of younger rock lying beneath older ones. Those travelers who climb Gebel Musa (Mt. Sinai) in the dark to enjoy the sunrise on its summit will not be able to admire the rocks on the way up, but in the daylight the reds, roses, and grays of the peaks cross-hatched with their black and red dikes is very dramatic. Fortunately, 4,350 sq km of this beautiful but fragile area was designated as the St. Katherine Natural Protectorate in 1996.

Today, Bedouins are the chief inhabitants of the interior of the Sinai Peninsula. They make a living by combining agriculture (palm groves and orchards near a source of permanent water) and animal husbandry (sheep, goats, and camels, which are taken to various areas for pasture). Some live on the Wadi el-Arish 'floodplain,' where they traditionally planted barley and melons that were watered by the winter rains. They are increasingly turning to more sedentary lives and work in the mining, oil, or tourism industries. They have also been encouraged to grow olives, fruit, and vegetables for the El-Arish markets. Some are employed as guards in the national protectorates, where they use their knowledge of the wilderness to preserve their ancient homeland.

The Egyptian Government has made development of Sinai a national priority. As a result, the population of Sinai has increased dramatically in the last few decades, primarily by immigration from the crowded Nile-Valley and Delta. The Egyptian Army has been used to help build infrastructure. This has included constructing roads, bridges, a pipeline carrying fresh water from Suez down the west coast to Sharm el-Sheikh, factories to process agricultural products, and hospitals.

Several agricultural reclamation schemes have been proposed or are underway to bring fresh or brackish irrigation water from the Nile to areas east of the Suez Canal and along the Sinai Mediterranean coast. The existing Ismailiya Canal is being extended to irrigate land east of the Suez Canal, while a new canal, the Salam (Peace) Canal, has been constructed to provide irrigation water for land south of Lake Manzala and along the Mediterranean coast as far as El-Arish. An 1,800 sq km agricultural project is planned for the alluvial soil in Wadi Girafi in east-central Sinai. It will employ a variety of water management schemes, including deep wells, canals, reservoirs, and diversion dams.

Conclusions

In this short book I have provided an introduction to the main principles of geology, along with a summary of the geological history of Egypt in order to show how that history shaped the environment in which Egyptian civilization arose and thrived for over five thousand years. A few main themes have been highlighted and explored in many of the chapters. As we recap these ideas, we can also ask: What about the future? Will the narrow Nile Valley that served as the 'cradle' for Egypt's spectacular past continue to provide, or will it become a straitjacket hindering progress and threatening the quality of life?

We have examined the resources that Egypt's geological past provided for its inhabitants. The largest resource is rocks of many useful types—located in places where they can be quarried conveniently and used for building, tools, and art. The ancients used limestone, sandstone, and granite for their enduring monuments. Everyday dwellings were made of perishable but renewable mud bricks. Even the Romans appreciated the beautiful stone available in Egypt and quarried columns, paving blocks, and material for statuary in their European cities, as well as carrying off finished objects like obelisks.

Modern Egyptians use many of the ancient quarries to mine limestone for production of cement. They have replaced their mud brick houses with ones of poured concrete and fired clay bricks. Gravel and sand have become important materials for road building and industrial processes. Thus the geological features continue to shape the modern society, albeit in a different manner than in the past.

For the ancients, up to and including the Greeks and Romans, Egypt's mineral resources were vital assets. Gold, copper, gems, and pigments were among the most prized. Today, minerals such as phosphates, iron and manganese ores, and gypsum are exploited in huge quantities. The Egyptians are also fortunate in having reserves of oil and natural gas that allow them to be self-sufficient as well as to export these fuels. Hydroelectric power from the Aswan High Dam generating station supplements these energy sources, as Egypt builds an industrial base.

Egypt today faces two geological limitations that are not encouraging and that cannot be ignored: the amount of cultivatable land and the supply of water. Egypt's agricultural practices have made full use of the land of the Nile Valley and the Delta, which together total around 35,000 sq km. Most of the area is perennially irrigated and produces two or three crops annually. Given a population of 70 million in 2001 CE, this works out to two thousand people supported per square kilometer. This ratio of people to land area has been increasing as a result of population growth that now averages over a million people a year. Land reclamation has not kept pace with the population growth, and agricultural land is still being lost to urbanization, despite government edicts.

Numerous schemes to use desert land for cultivation have been mentioned in this book, and many more will be proposed in the future. According to Egyptian officials, 25 percent more land was being farmed in 1999 than in 1952 as a consequence of water supplied by the High Dam. The success of reclamation projects has been hard to determine, however. Most lands targeted, while the best candidates for development, nonetheless had poor soil potential. The cost of putting an acre into cultivation has been incredibly high, and the average increases as greater challenges are attempted. The final limit on reclamation will be the availability of water, however, not land.

We have seen that historically, Egypt has been almost totally reliant on the River Nile. Except for some wells in the mountains and oases and a small amount of winter rain along the Mediterranean coast, all water for crops, people, and industry came from the Nile. We saw how ancient Egyptians learned to make effective use of the annual inundation for their single winter crop. The Ptolemies introduced several water-lifting devices that made it possible to grow two crops on some raised areas. They made the export of grain an important part of their economy; a practice continued by the Romans, when Egypt was the 'breadbasket of the Empire.'

Not until the nineteenth century CE did Egyptians find a way of saving some of the Nile's annual floodwaters to supply summer crops. Then cotton and sugar cane became major cash crops, raised for export rather than to feed the local population. The Aswan High Dam was designed to make perennial irrigation possible the entire length of the Nile Valley and in the Delta. But even as agricultural productivity rose, the population increased faster. Today, Egypt imports about half of its wheat. It could grow more food for home consumption, but fruits and vegetables for export or for city dwellers bring a higher price to the individual farmer.

Egypt's allotment of Nile water is set by international treaty at 55.5 billion cubic meters per year. This value was determined at a time when Egypt and Sudan were the main users of Nile water. Other riparian countries received enough rainfall for their needs. Today, upstream countries such as Ethiopia, Uganda, and Kenya are planning to increase their use of Nile water in ways that may affect Egypt's portion. Egypt can only hope to maintain its historical share and attempt to use the water more effectively. Already it derives another 15 billion cubic meters for use by recycling and deep wells.

Since agriculture currently uses 85 percent of the water, that is where efficiencies would realize the most benefit—efficiencies such as watertight distribution canals and waterpipes, drip or fixed irrigation instead of flooding basins, and the use of drought-tolerant crops instead of water-hungry cotton, sugar cane, and rice. The extent to which the Nubian aquifer of the Western Desert and deep reserves in Sinai can be exploited is still open to question. Desalinization is probably not a solution for agricultural purposes, although it can meet the needs of some towns in especially arid locations.

We have learned that the landscape of Egypt is the product of geological processes that began in the distant past and are continuing into the present. Many features, however, were initiated during a much wetter period and cannot be explained simply by reference to today's arid climate. Many of these geological processes are so slow that we cannot observe them in action: continental drift, uplift of mountains, or erosion of highlands. Then an earthquake or volcanic eruption reminds us that these forces are indeed at work. Other processes act at rates we can measure in our lifetime: weathering of rocks and erosion of soils can occur rapidly under the right conditions. And major changes in the water table and soil salinity can result from building dams. Geology can provide the data and the perspective to recognize conditions operating in the past that may not be obvious today. Geology can offer even the nonspecialist a better understanding of the environmental factors that shape not only Egypt but also every human society.

Acknowledgments

A ll the sources that I consulted during the preparation of this book are list-ed in the Bibliography. I am especially indebted to publications by Dr. Rushdi Said, former director-general of Egypt's Geological Survey; Dr. Bahay Issawi, former vice-minister, Egypt's Ministry of Petroleum and Mineral Wealth; Dr. Karl Butzer, professor in the Departments of Geography and Anthropology, University of Texas in Austin; and Dr. James Harrell, professor of geology, University of Toledo, Toledo, Ohio. I am pleased that Dr. Issawi gave me permission to reproduce three figures from his article, "The Cenozoic Rivers of Egypt: The Nile Problem." I have benefitted from discussions with Dr. Issawi and with Dr. Harrell, whose suggestions on the manuscript were very much appreciated.

I am grateful for access to materials and assistance from the Chapel Hill Public Library and its Interlibrary Loan Department; the Libraries of the University of North Carolina at Chapel Hill; the Libraries of Duke University; the Wilbour Library of The Brooklyn Museum of Art; the American Research Center in Egypt's Library in Cairo; the Library of the American University in Cairo; and the Library at the Egyptian Geological Survey and Mining Authority.

I want to thank Dr. Zahi Hawass, Secretary General of the Supreme Council of Antiquities, for his support in visiting some of the sites. I have enjoyed the friendship and guidance of Egyptologists Mohammed A. Shata and Mansour Bouriak Radwan and guides Mahmoud Khodair and Mohamed Rezik. They have been tireless in showing me around and sharing their knowledge and love of ancient Egypt.

I am happy to include the excellent color photographs provided by Elaine Godwin (EG), Richard Harwood (RH), Mahmoud Khodair (MK), Frank

Pettee (FP), Charles Rigano (CR), and R. Bruce Sampsell (RS). Photographs by these individuals are identified with their initials at the end of the captions. Photographs without any initials were taken by me. The three satellite images came from Jacques Descloitres, MODIS Land Science Team and can be found under Egypt for February 28, 2000 at http://visibleearth.nasa.gov.

Many friends have encouraged me in this endeavor and helped in various ways. My everlasting thanks go to each of them. I am deeply grateful for the loving support and able technical and editing assistance of my husband, Bruce Sampsell, during all phases of this project.

Figure Credits

A ll figures in this book were drawn by the author. The following published sources were consulted in their production.

Fig. 1.1. Buckle, 1978; Birkeland and Larson, 1989.
Fig. 1.2. Birkeland and Larson, 1989.
Fig. 1.3. Birkeland and Larson, 1989.
Fig. 2.1. Winkler, 1994.
Fig. 2.2. Ball, 1939; Said, 1962; Klitzsch, List, and Pohlmann, 1986–87; Said, 1990.
Fig. 4.1. Said, 1993.
Fig. 4.2. Said, 1981.
Fig. 5.1. Space photograph taken on June 28, 2001 and available at http://oel.jsc.nasa.gov/scripts/sseop/photo.pl?mission=ISS002&roll=ESC&frame=7744.
Fig. 5.2. Pritchard, 1979.
Fig. 6.1. Ball, 1907; Aston, Harrell, and Shaw, 2000.
Fig. 6.2. Buckle, 1978.
Fig. 7.1. Sandford and Arkell, 1933; Ball, 1939; Said, 1962.
Fig. 9.1. Klitzsch, List, and Pohlmann, 1986–87.
Fig. 9.2. Klitzsch, List, and Pohlmann, 1986–87.
Fig. 9.3. Caton-Thompson and Gardner, 1934; Ball, 1939; Harrell and Bown, 1995; Wadi Rayan lakes based on space photograph taken on February 26, 2001 and available at http://eol.jsa.nasa.gov/scripts/sseop/photo.pl?mission=ISSOO2&roll=701&frame=147.
Fig. 10.1. Klitzsch, List, and Pohlmann, 1986–87.
Fig. 10.2. Kemp, 1989.

Fig 11.1. Ball, 1942; Said, 1981.

Fig. 11.2. Said, 1981; Greenwood, 1997.

Fig. 12.1. Strabo, 1932 edition; Ball, 1942; Frazier, 1972; Empereur, 1998.

Fig. 12.2. Strabo, 1932 edition; Frazier, 1972; Empereur, 1998.

Fig. 13.1. Said, 1990.

Fig. 13.2. Issawi and McCauley, 1992. Reproduced with permission of Dr. Bahay Issawi.

Fig. 13.3. Issawi and McCauley, 1992. Reproduced with permission of Dr. Bahay Issawi.

Fig. 13.4. Issawi and McCauley, 1992. Reproduced with permission of Dr. Bahay Issawi.

Fig. 14.1. Hume, 1935; Klitzsch, List, and Pohlmann, 1986–87.

Fig. 14.2. Murray, 1925; Hume, 1935, 1937; Klitzsch, List, and Pohlmann, 1986–87; Sidebotham and Wendrich, 1999.

Fig. 15.1. Martinez and Cochran, 1988.

Fig. 15.2. Klitzsch, List, and Pohlmann, 1986–87.

Fig. 16.1. Drawing by James Wyld, 1869, published in Kinross, 1969.

Fig. 17.1. Rothenberg, 1979; Greenwood, 1997.

Fig. 17.2. Greenwood, 1997.

Glossary

aeolian (also **eolian**): referring to the wind.

aggrade: to build up the bed of a river or the surface of the floodplain by depositing sediments.

alluvial fan: material deposited by a stream when it enters a plain; the decrease in gradient reduces the water's ability to carry sediment.

alluvium: sediments deposited by flowing water.

aquifer: a porous layer of rock, sand, or gravel that holds water.

aragonite: a form of calcium carbonate produced by mollusks and certain other marine organisms that contribute to coral reefs. After the death of such organisms, the aragonite can recrystalize as **calcite**.

architrave: a horizontal beam spanning the space between two columns.

artesian well: a well that penetrates into an **aquifer** in which the water is under pressure because it is confined between two impermeable layers of rock. Water will rise in the well bore due to this pressure—it may or may not flow out at the surface. In a spring, by contrast, the water table intercepts a depression so that the water flows out onto the surface.

barrage: a dam whose function is to raise the level of water in a section of a river; this allows water to be diverted into an irrigation channel.

basalt: a very dense, dark, fine-grained extrusive igneous rock; the most abundant volcanic rock worldwide.

base level: the level to which erosion will reduce a landmass. Globally, sea level is the base level, but locally it may take the form of a river channel.

basement complex: in Egypt, the oldest layer of rock, composed of ancient igneous and metamorphic rock types. Labeled 'BC' on maps when exposed at the surface.

basin: a depression toward which streams flow. It collects sediments.

bedding plane: surface between two beds or **strata** in a sedimentary rock, usually horizontal at the time sediments are deposited. The plane represents a break in the deposition process or a change in conditions.

bedrock: the solid rock exposed at the surface of the ground or that underlies soil and loose sediments.

bir: Arabic word for 'well.'

butte: a landform with a flat top and steep sides. It is a section of a plateau that has been isolated by erosion.

calcite: common form of calcium carbonate, the main mineral in limestone.

Canopic: westernmost branch of the Nile as it crossed the Delta in ancient times.

cataract: rapids in a river with many rocks emerging above the water.

catchment area: the area from which rain drains into a single river; the same as a **drainage basin**.

chalk: a fine-grained limestone formed of the skeletons of marine microorganisms.

clay: rock particles smaller than 0.004 millimeters. The term clay can also refer to several kinds of clay minerals that can absorb and lose water quickly and are plastic when wet and hard when dry. The reader should take care to determine which of these two meanings is intended in a particular context.

conglomerate: sedimentary rock type in which the particles consist of rounded rock fragments greater than 2 mm in size.

coprolite: fossilized feces.

craton: a section of continental crust that has not undergone rifting or mountain building for the last billion years. Cratons form the oldest parts of each continent, onto which additional crust has been added.

crust: outermost layer of the Earth. There is a dense, brittle layer about 5 km thick under the ocean basins; the continents are made of a lighter material, and the crust beneath mountain ranges is up to 60 km thick. Compare with **lithosphere**.

Damietta: one of the branches of the River Nile as it flows across the Delta and enters the Mediterranean near the coastal town of Damietta. Today, it is the eastern of the two existing branches.

deflate: to remove material from a surface by wind action.

degrade: to lower the bed of a river or other surface.

denude: to remove material from the landscape through processes of weathering, erosion, and mass wasting.

dike (also **dyke**): a thin seam of igneous rock forced up through a fissure in pre-existing rock; a raised ridge to channel water.

distributary: one of the branches into which a single river channel divides, especially as it crosses a delta or region of low gradient; the opposite of a tributary.

dolerite: a dense, fine- to medium-grained, dark intrusive igneous rock formed in shallow intrusions such as **dikes**. Ancient Egyptians used balls of dolerite as pounders for working granite.

drainage basin: the entire area from which rainfall collects and flows into one river system.

Egyptian alabaster: a form of travertine, or calcium carbonate, similar in appearance to real alabaster (calcium sulfate or gypsum) and used in Egypt for statuary, vessels, pavement, and offering tables.

Eon: the largest time interval in the Geological Time Scale.

Eonile: the earliest major north-flowing river in Egypt formed in the Miocene. During the late-Miocene desiccation of the Mediterranean Sea, this river cut the Nile Canyon, which defines the modern Nile Valley. See Table 3.2.

Epoch: a subdivision of a **Period** in the Geological Time Scale.

Era: a subdivision of an **Eon** in the Geological Time Scale; subdivided into **Periods**.

erosion: removal and transport of weathered rock by wind, running water, waves, ice, or the force of gravity.

escarpment: a cliff or steep slope bordering a plain.

evaporite: rock (like gypsum, anhydrite, or rock salt) formed by precipitation of minerals when water evaporates.

extrusive: referring to **igneous** rocks that cool above ground.

fault: a crack in a rock mass in which one side has moved relative to the other; movement may be up, down, or sliding past.

fault block: a section of crust delimited by faults.

feeder canal: a canal that carries irrigation water from the river to a basin or field.

flint: fine-grained rock formed within limestone strata by the precipitation of silica (silicon dioxide or quartz); a type of chert. Used for making stone tools (axes, scrapers) and arrowheads.

floodplain: the flat area on each side of a river that is covered with water when the river is in its flood stage.

formation: a mass of rock with a distinct set of geological properties; several formations can be combined into a **group**; subdivisions of formations are called members.

frit: a synthetic, glass-like material containing copper ores that was powdered to produce blue and green pigments for paints.

gebel: Arabic word for 'mountain.'

gezira: Arabic word for 'island.'

gneiss: a metamorphic rock with compositional layering usually derived from igneous rocks such as granite, diorite, or gabbro.

Gondwana: a former land mass composed of the modern continents of Africa, Antarctica, Australia, and South America. It began to split up around 200 million years ago as part of the breakup of **Pangaea**.

gradient: the slope of a stream from its headwaters to its **base level**; the greater the gradient, the more erosive the stream. Note: gradients are expressed as ratios, e.g., 1:200, while slopes are expressed in degrees.

granite: a coarse-grained intrusive igneous rock, generally light in color or pink or reddish; used extensively by ancient Egyptians as a building material and for sarcophagi, obelisks, and statuary. Grayish or black 'granite' is actually the igneous rock granodiorite.

ground water level: the level below which the ground is saturated with water; same as water table.

group: a category composed of several rock **formations**.

gypsum: an **evaporite** of hydrous calcium sulfate ($CaSO_4*2H_2O$). Lumps of gypsum crystals occur naturally among limestone outcroppings throughout Egypt. This was used throughout pharaonic times for plaster (today called plaster of Paris) and mortar. Gypsum plaster was often used as a finish coat over one or more coats of mud/clay plaster on walls. Such a surface could be carved or painted. No example of lime mortar or cement has been found in Egypt that predates the Greek and Roman eras. This is probably due to the shortage of fuel sufficient to prepare lime cement, since the raw material, limestone, was certainly not in short supply. To produce cement from either gypsum or lime, it is first necessary to pulverize the rock and then to heat it in order to drive off some of the bound water, in a process known as calcining. Production of quicklime, the raw ingredient for cement, requires heating well-pulverized limestone to 900° C (1,652° F). Raising large quantities of limestone to this temperature and holding it there for a sufficiently long time was beyond the capacities of the dynastic Egyptians. Additional (human) energy would have been needed to grind the rock to a fine powder afterward. On the other hand, ground gypsum need only be heated to 130° C (268° F), at which temperature it will form a material that has the ability to recombine with water to form a paste.

hematite (also haematite): red mineral composed of iron oxide; used as an iron ore or a pigment.

Herodotus: Greek historian who visited Egypt in about 450 BCE; he recorded his observations plus a lot of information he was given by priests—the former are considered useful, while the latter contains many questionable ideas.

Holocene Wet Phase: a period from 10,000 to 5,000 years before the present, during which southern Egypt received a greater annual rainfall than it does today, leading to a higher than usual flood level in the Nile. The Western Desert also received more rain and many areas were inhabited either perennially or seasonally.

igneous: any kind of rock that forms from cooling **magma**. Along with **sedimentary** and **metamorphic** rocks this is one of three main categories of rock types.

index fossil: a fossil whose presence can be used to date a rock.

intrusive: referring to **igneous** rocks that are intruded into pre-existing rocks; because they cool slowly below ground they have larger crystals than rocks that cool rapidly on the surface.

isostasy: a process of adjustment in the elevation of crustal blocks. Because each crustal block is essentially floating on the semi-molten mantle beneath it, thicker blocks both have a higher elevation above sea level and extend deeper into the mantle than thin blocks. If a thick block loses material from its upper surface by erosion, the resulting thinned block will float higher than before, paradoxically raising its surface.

joint: one of a series of similarly oriented cracks or fractures in a rock along which no movement of the two parts of the rocks has occurred.

lava: molten rock (**magma**) that erupts from a volcano or fissures.

leaning vault: brick vault formed by setting a few bricks against an end wall perpendicular to the two sidewalls to be joined by the vault. The next course of brick leans against the first and reaches a bit higher. This continues until a course reaches high enough to close at the top of the vault. From the side, the courses have a visible slant toward the end wall. The advantage of this type of vault is that it can be built without the use of wooden centering.

levee: a raised ridge parallel to the course of a river. Formed naturally on both sides of a river when it floods its banks and drops sediments along the bank; it can also be made artificially to control flooding of the river or to direct irrigation waters.

lithify: to change sediments into a rock by cementing the grains or rock fragments together.

lithosphere: the solid outer layer of the Earth; it is composed of the **crust** and the upper portion of the **mantle** that is cooler and more brittle than the bulk of the mantle below. The lithosphere is about 60 km thick near mid-oceanic ridges, 130 km thick beneath older ocean basins, and up to 300 km thick beneath continental cratons. It is divided into tectonic plates.

magma: molten rock, often formed by melting of subducted crust.

mantle: layer of Earth beneath the crust and above the core; it is about 2,300 km thick, hot, and at least partially molten. The uppermost layer of the mantle is cooler and more brittle; together with the **crust**, this uppermost layer forms the **lithosphere**.

marl: a soft sedimentary rock containing calcium carbonate (comprising from 35 to 65 percent of total volume) and clay mineral particles.

massif: a major mass of rock composing a mountain.

meander: a loop of a river.

metamorphic: category of rocks that have been changed from their original form into another kind by heat and pressure. Both **sedimentary** and **igneous** rocks can be metamorphosed.

metric unit equivalents:

1 kilometer (km) = 0.6 mile

1 meter (m) = 3.28 feet

1 centimeter (cm) = 10 millimeters (mm) = 0.4 inches

1 square kilometer (sq km) = 0.386 square mile = 247 acres

1 billion cubic meters = 0.24 cubic miles = 264 billion gallons

1 kilogram (kg) = 2.2 pounds

1 metric ton = 1.1 English ton

Mokattam: named for a hill east of Cairo (Gebel el-Muqattam). This name was originally applied to a formation composed of thick dense limestone of Middle Eocene age. In the new *Geological Map of Egypt* published in 1986–87, the rock categories were modified, with Mokattam now referring to a group that contains a number of formations including the former Mokattam, now called the Observatory Formation. The term Mokattam Formation is retained in this book.

Neolithic: the New Stone Age. A period beginning around 8000 BCE when humans began to live in fixed villages, plant crops, herd animals, and make pottery and textiles. Tools were made of polished stone; compare with **Paleolithic**.

Neonile: one of the several phases that the River Nile has experienced. It flowed during the Pleistocene from about 400,000 to 12,500 years before the present. See Table 3.2.

Nilometer: a structure built on the bank of the Nile with a vertical scale to measure the height of the river, especially while in flood.

Nubia: ancient country south of Aswan; composed today of southern Egypt and northern Sudan.

Nubian sandstone: a rock layer formed in the Cretaceous; it covers much of southern Egypt and northern Sudan. It is been divided into a number of separate formations, which are ignored in this book. Labeled 'NS' on maps when exposed at the surface.

nummulitic limestone: rock containing fossils of single-celled organisms belonging to the genus *Nummulites*. There were many species of this widespread and rapidly evolving genus during the Eocene Epoch, which make them useful for dating and correlating rock units. *Nummulites gizehensis* had a disc-shaped, calcium carbonate test (shell), nearly 2 cm in diameter. These organisms belong to the order *Foraminifera*.

oasis: an area in an otherwise arid landscape where water is available year round allowing the growth of some vegetation and potential human habitation.

oolitic limestone: a type of limestone formed mainly of small, sand-sized grains of calcium carbonate called ooliths (egg-shaped). True ooliths are composed of many thin concentric layers of calcium carbonate; they form only in shallow seas without land-derived sediments. Their spherical or egg shape is the result of tumbling in highly saturated seawater. In the ridges of limestone along the northwest Mediterranean coast of Egypt, many of the sand-sized grains are simply small pieces of waterworn seashells and not true ooliths. A geologist would call such a limestone 'bioclastic oolitic calcarenite.'

outcrop: part of a rock formation exposed at the Earth's surface.

Paleolithic: Old Stone Age; a period of human technical development characterized by stone tools. Three divisions with an increasing complexity of stone implements have been defined: Lower Paleolithic with pebble tools and bifacial hand axes, Middle Paleolithic with flaked tools, and Upper Paleolithic with blade-based tools of greater variety and complexity as well as tools containing many small pieces of stone or microliths. The dates of the three divisions are approximately: before 60,000 years ago, 60,000 to 40,000 years ago, and 40,000 to 10,000 years ago.

Paleonile: one of the phases of the Nile; the first river flowing through the entire country of Egypt from south to north. It flowed during the late-Pliocene Epoch replacing an arm of the Mediterranean that filled the Nile Canyon during the early Pliocene. See Table 3.2.

Pangaea: a supercontinent containing all the Earth's land masses; in existence from around 290 to 250 million years ago.

pediplain: a landscape resulting from the erosion of a planar land mass to a base level, often forming plains at several different elevations; the uneroded remnants have steep **escarpments**. It is characteristic of arid regions, but probably began its development during an earlier, wetter period.

Pelusiac: one of the ancient branches of the Nile across the Delta; it was the easternmost branch, leading toward Sinai.

perennial irrigation: a system that supplies irrigation water on a year-round basis so that several crops can be grown on the same field, in contrast to basin irrigation, in which the annual flood supported a single crop per year.

Period: One of the subdivisions of an **Era** in the Geological Time Scale; further subdivided into **Epochs**.

pirated: the capture by one river system of a tributary of another river system; this occurs when one river erodes 'headward' or upstream through a barrier or divide that formerly separated the two different drainage basins. The river with the lower **base level** or steeper gradient will capture the other one.

plate tectonics: a unifying theory in geology in which the main features are the division of the Earth's crust and upper mantle (called the **lithosphere**) into a number of plates, the movement of the plates causing the formation of new oceanic crust from upwelling magma at a divergence zone and the loss of oceanic crust by subduction at a convergence zone.

playa: an inland depression into which water drains from surrounding highlands forming an ephemeral lake; sediments include **clays** and **evaporites**.

pluton: a body of intrusive igneous rock.

pluvial: a period of greater than normal rainfall.

porphyry: an intrusive igneous rock in which visible crystals constituting more than 20 percent of the rock's volume are dispersed through a homogeneous finer-grained groundmass.

Prenile: one of the phases of the River Nile; in existence for several hundred thousand years during the Pleistocene. A vigorous river of high volume, it carried and deposited great quantities of sediment. See Table 3.2.

pressure release: a process that causes **joints** to form in igneous rocks that have cooled deep in the Earth under great pressure when they are uplifted and/or their overburden is eroded away.

primary producer: any organism that can synthesize its own complex organic molecules, generally by using sunlight (photosynthesis) as its source of energy.

Punt: a region along the southern Red Sea that was a destination for ancient Egyptian trading expeditions. Its exact location has been debated but may correspond to modern Sudan, Eritrea, or Somalia.

quartzite: a quartz-cemented (siliceous) sandstone that is much harder than other varieties of sandstone. The term quartzite is also applied to metamorphosed sandstone of any composition.

questa: a landform in which a tilted block of rock has a steep **escarpment** on one side and a gradual slope on the other.

radar rivers: an extinct river system in the southwestern Western Desert, whose channels are now filled with sediment and hidden by sand. It was detected by radar carried by satellites.

ras: Arabic word for 'headland' or 'cape.'

rejuvenate: to increase the gradient of a stream either by uplifting its catchment area or lowering its base level, which can be sea level or the point where the stream enters a river.

rift valley: a valley formed when crustal stretching causes faults that permit a crustal block to drop down between two neighboring regions.

rifting: breaking apart of a lithospheric plate.

Rosetta: one of the branches of the Nile flowing across the Delta; today it is the western branch and meets the Mediterranean at Rosetta.

sabakh: mud from old mud-brick walls or buildings that is sought after by farmers as a fertilizer for their fields because it contains high concentrations of nitrates.

sabkha: a lagoon or salt marsh.

sand: rock or mineral fragments whose particle size is between 0.0625 and 2 millimeters.

saqya: a device for lifting water that consists of a vertical water wheel (or chain of pots) geared to a horizontal wheel that can be turned by a person or an animal—usually a donkey or cow; introduced into Egypt during the Ptolemaic period. Another water-raising device, the Archimedes screw, was introduced in the fifth century BCE.

scarp: variant of **escarpment**.

scree: loose rock fragments forming an inclined surface at the base of a cliff from which they have fallen; same as talus.

sediment: material eroded from rocks that is removed from one location and deposited at another.

sedimentary: kind of rock that is formed by the accumulation of rock fragments, minerals, or biological remains deposited in an inland or ocean basin; fragments are cemented by minerals that precipitate from seawater or ground water.

shadoof: a device for raising water consisting of a long pole pivoted on a post that has a bucket at one end and a counterbalancing weight at the other. It is worked by hand by lowering the bucket into the water source, then the counterweight lifts the bucket so the water can be emptied out at a higher level.

shale: a fine-grained type of sedimentary rock containing particles of silt or clay.

silt: rock particles in size range of 0.004 to 0.0625 millimeter.

sluice gate: a gate in a channel that controls the flow of water.

solution weathering: a type of chemical weathering in which weak carbonic acid (formed when rain absorbs carbon dioxide from the atmosphere) dissolves calcium carbonate in limestone (or other rock), leaving voids that form fissures, tunnels, and caves.

spall: a flake or chip of stone; in this book, one removed by a natural weathering process (referred to as spalling).

stela: an inscribed stone put up to commemorate an event or as a funerary marker.

stratum (plural **strata**): a layer of sedimentary rock with a discernable set of properties.

stratigraphy: the study of the relationship of rock layers or other deposits as they are formed one above another.

subduction: process of one crustal plate diving beneath another plate at a convergence zone. The subducting crust melts as it enters the upper mantle of the Earth.

tectonic: refers to deformations in the Earth's crust such as folding, uplift, or faulting resulting from magma upwelling or plate movements such as collisions.

Tertiary: first Period of the Cenozoic Era, lasting from around 65 million years ago to 2 million years ago; comprises Epochs of Paleocene, Eocene, Oligocene, Miocene, and Pliocene. A very important time interval for the formation of the Egyptian landscape, during which the Red Sea opened, the Red Sea Mountains were uplifted and eroded, the Western Desert was denuded, depressions formed, and the River Nile came into being.

Tethys Sea: the ocean that once separated the land masses of Africa and Eurasia; as these two plates converged the Tethys Sea floor was subducted; the Mediterranean is the modern remains of this sea. This sea—lying north of Egypt—repeatedly submerged the land, permitting the formation of many layers of sedimentary rock.

turtleback: a ridge of sand deposited by the Prenile river on the Delta during the Pleistocene that has resisted erosion and rises above the surface of later deposits of silt.

volcanic: processes or products produced in association with volcanoes or fissures through which **magma** emerges onto the surface of the Earth.

wadi: a streambed with an ephemeral flow of water; usually dry with steep walls and much alluvium in its bed. In the southwestern United States it is called an *arroyo*.

watershed: In North America, this term is a synonym for **drainage basin**; in Britain the term refers to the divide between two adjacent drainage basins.

weathering: mechanical or chemical breakdown of rock into smaller pieces or chemical alteration into different minerals.

yardangs: streamlined, wind-sculpted rocks in the shape of a resting camel.

Bibliography

Abdel-Wahab, H.S., K. Yemane, and R. Giegengack. 1997. "Mineralogy and geochemistry of Pleistocene lacustrine beds in Wadi Feiran, South Sinai, Egypt: Implications for environmental and climate changes." *Egyptian Journal of Geology* 41 (2A):145–72.

Abu Al-Izz, M.S. 1971. *Landforms of Egypt*. Cairo: The American University in Cairo Press.

Allaby, Ailsa and Michael Allaby, eds. 1991. *The Concise Oxford Dictionary of Earth Sciences*. Oxford and New York: Oxford University Press.

Ambraseys, N.N., C.P. Melville, and R.D. Adams. 1994. *The Seismicity of Egypt, Arabia and the Red Sea*. Cambridge: Cambridge University Press.

Arnold, Dieter. 1991. *Building in Egypt: Pharaonic Stone Masonry*. Oxford and New York: Oxford University Press.

Aston, Barbara G., James A. Harrell, and Ian Shaw. 2000. "Stone." In *Ancient Egyptian Materials and Technology*. Paul T. Nicholson and Ian Shaw, eds. Cambridge: Cambridge University Press.

Baines, John and Jaromir Malek. 2002. *Atlas of Ancient Egypt*. Revised Edition. Cairo: The American University in Cairo Press.

Ball, John. 1907. *A Description of the First or Aswan Cataract of the Nile*. Cairo: National Printing Department.

———. 1939. *Contributions to the Geography of Egypt*. Cairo: Government Press.

———. 1942. *Egypt in the Classical Geographers*. Cairo: Government Press, Bulaq.

Bell, Barbara. 1971. "The Dark Ages in Ancient History: I. The First Dark Age in Egypt." *American Journal of Archaeology* 75:1–16.

Bietak, Manfred. 1996. *Avaris, the Capital of the Hyksos: Recent Excavations at Tell el-Daba.* London: British Museum Press.

Birkeland, Peter W. and Edwin E. Larson. 1989. *Putnam's Geology.* 5th Edition. Oxford and New York: Oxford University Press.

Bowman, Alan K. 1996. *Egypt After the Pharaohs.* Berkeley and Los Angeles: University of California Press.

Bowman, H., F.H. Stross, F. Asaro, R.L. Hay, R.F. Heizer, and H.V. Michel. 1984. "The Northern Colossus of Memnon: New Slants." *Archaeometry* 26(2):218–29.

Buckle, Colin. 1978. *Landforms in Africa.* London: Longman Group Ltd.

Butzer, Karl. 1960. "On the Pleistocene Shore Lines of Arabs' Gulf, Egypt." *Journal of Geology* 68:626–37.

———. 1976. *Early Hydraulic Civilization in Egypt.* Chicago: University of Chicago Press.

———. 1976. *Geomorphology from the Earth.* New York: Harper and Row.

———. 2001. "When the Desert was in Flood: Environmental History of the Giza Plateau." *AERAGRAM* (Newsletter of the Ancient Egypt Research Associates) 5(1):3–5.

Butzer, Karl W. and Carl L. Hansen. 1968. *Desert and River in Nubia.* Madison: University of Wisconsin Press.

Casson, Lionel. 1954. "The Grain Trade of the Hellenistic World." *Transactions of the American Philological Association* 45:168–87.

———. 1991. *The Ancient Mariners.* 2nd edition. Princeton: Princeton University Press.

Caton-Thompson, Gertrude and Elinor W. Gardner. 1934. *The Desert Fayum.* (2 vol.) London: Royal Anthropological Institute of Great Britain and Ireland.

Clark, Somers and R. Engelbach. *Ancient Egyptian Construction and Architecture.* 1930 as *Ancient Egyptian Masonry*; reprint, New York: Dover Publications, Inc., 1990.

Clayton, Peter A. 1994. *Chronicle of the Pharaohs.* London: Thames and Hudson.

Coleman, Robert G. 1993. *Geological Evolution of the Red Sea.* New York: Oxford University Press.

Craig, G.M., ed. 1993. *The Agriculture of Egypt.* Oxford: Oxford University Press.

Dietz, Robert S. and John C. Holden. 1970. "The Breakup of Pangaea." *Scientific American*, October 1970.

Eddy, Frank W. and Fred Wendorf. 1999. *An Archaeological Investigation of the Central Sinai, Egypt.* Boulder, Co: University Press of Colorado.

Edwards, Amelia. *A Thousand Mile up the Nile.* Facsimile of 1888 edition; reprint, London: Parkway Publishing, 1993.

Empereur, Jean-Yves. 1998. *Alexandria Rediscovered.* New York: George Braziller Publisher.

Fitzgerald, Percy. 1876. *The Great Canal at Suez.* Vol. 1. London: Tinsley Brothers.

Forster, E.M. *Alexandria: a history and a guide.* 1922 edition; reprint, London: Michael Haag, Ltd. 1982.

Fraser, P.M. 1972. *Ptolemaic Alexandria.* Oxford, England: Clarendon Press.

Garfunkel, Z. 1988. "Relation between continental rifting and uplifting: evidence from the Suez rift and northern Red Sea." *Tectonophysics* 150:33–49.

Gauri, K. Lal, John J. Sinai, and Jayanta K. Bandyopadhyay. 1995. "Geological Weathering and Its Implications on the Age of the Sphinx." *Geoarchaeology* 10(2):119–33.

Geological Map of Egypt. 1986–87. See Klitzsch, List, and Pohlmann, eds.

Gohary, Jocelyn. 1998. *Guide to the Nubian Monuments on Lake Nasser.* Cairo: The American University in Cairo Press.

Graves, William. 1975. "New Life for the Troubled Suez Canal." *National Geographic* 147(6):792–817.

Greenwood, Ned H. 1997. *The Sinai: A Physical Geography.* Austin: University of Texas Press.

Habachi, Labib. 1965. "The graffito of Bak and Men at Aswan and a second graffito close by showing Akhenaten before the Hawk-headed Aten." *Mitteilungen des Deutschen Archäologischen Instituts Abteilung Kairo* 20:85–92.

Hamblin, Russell. D. 1988. "The Geology of the Gebel El-Rus Area and Archaeology Sites in the Eastern Fayum, Egypt." In *Excavations at Seila, Egypt.* C. Wilfred Griggs, ed. Provo, Utah: Religious Studies Center, Brigham Young University.

Harrell, James A. 1992. "Ancient Egyptian Limestone Quarries: A Petrological Survey." *Archaeometry* 34:195–211.

————. 1994. "The Sphinx Controversy: Another Look at the Geological Evidence." *KMT* 5(2):70–74.

Harrell, James A. and Thomas M. Bown. 1995. "An Old Kingdom Basalt Quarry at Widan el-Faras and the Quarry Road to Lake Moeris." *Journal of the American Research Center in Egypt* 32:71–91.

Hawass, Zahi. 1994. "Can the Sphinx be Saved?" *Archaeology* 47(5):42–43.

————. 1998. *The Secrets of the Sphinx: Restoration Past and Present.* Cairo: The American University in Cairo Press.

Hawass, Zahi and Mark Lehner. 1994. "The Great Sphinx of Giza: Who Built It and Why?" *Archaeology* 47(5):30–41.

————. 1994. "The Great Sphinx of Giza: Remnant of a Lost Civilization?" *Archaeology* 47(5):44–47.

Haynes, C. Vance. 1980. "Geochronology of Wadi Tushka: Lost Tributary of the Nile." *Science* 210:68–71.

Herodotus. *The Histories.* (Trans. Aubrey de Selincourt, with revised introductory matter and notes by John Marincola, 1996) London: Penguin Books.

Hewison, R. Neil. 2001. *The Fayoum: History and Guide.* Cairo: The American University in Cairo Press.

Hobbs, Joseph J. 1995. *Mount Sinai.* Austin: University of Texas Press.

Hoffmeier, James. 1993. "The Use of Basalt in Floors of Old Kingdom Pyramid Temples." *Journal of the American Research Center in Egypt* 30:117–23.

Holladay, John. S., Jr. 1982. *Cities of the Delta, Part III: Tell el-Maskhuta.* Malibu, CA: Undena Publications.

Hume, W. F. 1925. *Geology of Egypt. Vol. 1. The Surface Features of Egypt: their Determining Causes and Relation to Geological Structure.* Cairo: Government Press.

————. 1934. *Geology of Egypt. Vol. 2. The Fundamental Pre-Cambrian Rocks of Egypt and the Sudan; their Distribution, Age, and Character. Part I. The Metamorphic Rocks.* Cairo: Government Press.

————. 1935. *Geology of Egypt. Vol. 2. The Fundamental Pre-Cambrian Rocks of Egypt and the Sudan; their Distribution, Age and Character. Part II. The Later Plutonic and Minor Intrusive Rocks.* Cairo: Government Press.

————. 1937. *Geology of Egypt. Vol. 2. The Fundamental Pre-Cambrian Rocks of Egypt and the Sudan; their Distribution, Age and Character. Part III. The Minerals of Economic Value.* Cairo: Government Press.

————. 1962. *Geology of Egypt. Vol. 3. The Stratigraphical History of Egypt. Part I: From the Close of the Pre-Cambrian Episodes to the End of the Cretaceous Period.* Cairo: General Organization for Government Printing Offices.

————. 1965. *Geology of Egypt. Vol. 3. The Stratigraphical History of Egypt. Part II: From the Close of the Cretaceous Period to the End of the Oligocene.* Cairo: General Organization for Government Printing Offices.

Hsu, Kenneth. 1972. "When the Mediterranean Dried Up." *Scientific American* 227(6):26–36.

Issawi, Bahay and Mohamed el Hinnawi. 1980. "Contribution to the Geology of the Plain West of the Nile between Aswan and Kom Ombo." In *Loaves and Fishes: Prehistory of the Wadi Kubbaniya.* F. Wendorf and R. Schild (eds.) Dallas: Southern Methodist University.

Issawi, Bahay and John F. McCauley. 1992. "The Cenozoic Rivers of Egypt: The Nile Problem." In *The Followers of Horus,* Renee Friedman and Barbara Adams, eds. Egyptian Studies Association Publication No. 2. Oxbow Monograph 20. Oxford: Oxbow Books.

Kamil, Jill. 1993. *Aswan and Abu Simbel.* Cairo: The American University in Cairo Press.

Kearey, Philip, and Frederick J. Vine. 1990. *Global Tectonics.* London: Blackwell Scientific Publications.

Kemp, Barry J. 1989. *Anatomy of a Civilization.* London and New York: Routledge.

Kinross, John. 1969. *Between Two Seas: The Creation of the Suez Canal.* New York: William Morrow and Company, Inc.

Klemm, Rosemarie and Dietrich D. Klemm. 1993. *Steine und Steinbruche im Alten Ägypten.* Berlin and New York: Springer-Verlag.

Klitzsch, E., F.K. List, and G. Pohlmann, eds. 1986–87. *Geological Map of Egypt* (20 sheets, 1:500,000). Cairo: Conoco Inc. and Egyptian General Petroleum Corporation.

Larché, François. 1999. "The Reconstruction of the So-Called 'Red Chapel' of Hatshepsut and Thutmose III in the Open Air Museum at Karnak." *KMT* 10(4):56–65.

Lehner, Mark. 1994. "Notes and Photographs on the West-Schoch Sphinx Hypothesis." *KMT* 5(3):40–48.

————. 1997. *The Complete Pyramids.* London: Thames and Hudson.

Little, Tom. 1965. *High Dam at Aswan: the Subjugation of the Nile.* New York: John Day Company.

Lucas, A. and J. Harris. *Ancient Egyptian Materials and Industries.* 1962; reprint, London: Histories and Mysteries of Man Ltd. 1989.

Marlowe, John. 1964. *World Ditch: The Making of the Suez Canal.* New York: Macmillan Company.

Martinez, Fernando and James R. Cochran. 1988. "Structure and tectonics of the northern Red Sea: catching a continental margin between rifting and drifting." *Tectonophysics* 150:1–32.

El-Masry, N.N., B.A. El-Kalyouby, S.K. Khawasik, and M.A. El-Ghawaby. 1992. "Reconsideration of the geological evolution of Saint Catherine ring dyke, South Sinai." Pp 229–38 in *Proceedings of the Third Conference—Geology of the Sinai for Development.* Ismailia: Suez Canal University.

McDonald, John K. 1996. *The Tomb of Nefertari.* Cairo: The American University in Cairo Press.

McHugh, William P., John F. McCauley, C. Vance Haynes, Carol S. Breed, and Gerald P. Schaber. 1988. "Paleorivers and Geoarchaeology in the Southern Egyptian Sahara." *Geoarchaeology: An International Journal* 3(1):1–40.

McHugh, William P., Gerald G. Schaber, Carol S. Breed, and John F. McCauley. 1989. "Neolithic adaptation and the Holocene functioning of Tertiary palaeodrainages in southern Egypt and northern Sudan." *Antiquity* 63:320–326.

Murray, G.W. 1925. "The Roman Roads and Stations in the Eastern Desert of Egypt." *Journal of Egyptian Archaeology* 11:138–50.

Pawlicki, Franciszek and George B. Johnson. 1994. "Behind the Third Portico: Polish–Egyptian Restorers Continue Work on the Upper Terrace at Deir el-Bahri." *KMT* 5(2):41–49.

Peacock, David and Valerie Maxfield. 1994. "On the Trail of Imperial Porphyry." *Egyptian Archaeology* 5:24–26.

Penvenne, Laura J. 1966. "Disappearing Delta." *American Scientist* 84:438–39.

Phillips, James L. 1987. "Sinai During the Paleolithic: The Early Periods." In *Prehistory of Arid North Africa.* Angela E. Close, ed. Dallas, TX: Southern Methodist University Press.

Pritchard, J.M. 1979. *Landform and Landscape in Africa.* London: Edward Arnold.

Pryor, John. 1988. *Geography, Technology, and War: Studies in the maritime history of the Mediterranean, 649–1517 AD.* New York: Cambridge University Press.

Purser, Bruce. H. and Dan W.J. Bosence. 1998. *Sedimentation and Tectonics in Rift Basins: Red Sea, Gulf of Aden.* London: Chapman & Hall.

Redford, Donald B., ed. 2001. *The Oxford Encyclopedia of Ancient Egypt.* Oxford: Oxford University Press.

Reeves, Nicholas and Richard H. Wilkinson. 1996. *The Complete Valley of the Kings.* New York: Thames and Hudson.

Reilly, Frank A. 1964. *Guidebook to the Geology and Archaeology of Egypt.* Publisher not given.

Rogers, John J.W. 1966. "A History of the Continents in the Past Three Billion Years." *The Journal of Geology* 104:91–107.

Rogers, John J.W., Mohamed E. Dabbagh, Brian M. Whiting, and Sally A. Widman. 1989. "Subsidence and origin of the Northern Red Sea and the Gulf of Suez." *Journal of African Earth Sciences* 8:617–29.

Rothenberg, Beno. 1979. *Sinai: Pharaohs, Miners, Pilgrims and Soldiers.* Washington and New York: Joseph J. Binns.

Said, Rushdi. 1962. *The Geology of Egypt.* Amsterdam and New York: Elsevier Publishing Co.

————. 1981. *The Geological Evolution of the River Nile.* New York: Springer-Verlag.

————, ed. 1990. *The Geology of Egypt.* Rotterdam and Brookfield, VT: A.A. Balkema.

————. 1993. *The River Nile: Geology, Hydrology, and Utilization.* London: Pergamon Press.

Sandford, K.S. 1934. *Paleolithic Man and the Nile Valley in Upper and Middle Egypt.* Chicago: The University of Chicago Press.

Sandford, K.S. and W.J. Arkell. 1929. *Paleolithic Man and the Nile–Faiyum Divide.* Chicago: The University of Chicago Press.

————. 1933. *Paleolithic Man and the Nile Valley in Nubia and Upper Egypt.* Chicago: The University of Chicago Press.

Säve-Söderbergh, Torgny, ed. 1987. *Temples and Tombs of Ancient Nubia: The International Rescue Campaign at Abu Simbel, Philae, and Other Sites.* New York: Thames and Hudson.

Schoch, Robert M. 1992. "Redating the Great Sphinx of Giza." *KMT* 3(2):52–59, 66–70.

Selim, Amer A. 1974. "Origin and lithification of the Pleistocene carbonates of the Salum area, Western Coastal Plain of Egypt." *Journal of Sedimentary Petrology* 44(1):70–78.

Shaw, Ian and Paul Nicholson. 1995. *The Dictionary of Ancient Egypt.* New York: Harry N. Abrams, Inc.

Shukri, N.M., G. Philip, and R. Said. 1956. "The Geology of the Mediterranean Coast between Rosetta and Bardia. Part II. Pleistocene Sediments: Geomorphology and Microfacies." *Institut d'Egypte Bulletin* 37:395–433.

Sidebotham, Steven E. and Willemina Z. Wendrich., eds. 1998. *Berenike 1996: Report of the 1996 Excavations at Berenike (Egyptian Red Sea Coast) and the Survey of the Eastern Desert.* Leiden: Research School CNWS.

———. 1999. *Berenike 1997: Report of the 1997 Excavations at Berenike and the Survey of the Egyptian Eastern Desert, including Excavations at Shenshef.* Leiden: Research School CNWS.

Sorokin, Yuri. I. 1993. *Coral Reef Ecology.* Berlin: Springer-Verlag

Stanley, Daniel J. and Andrew G. Warne. 1993. "Nile Delta: Recent Geological Evolution and Human Impact." *Science* 260:628–34.

Stern, R.J. and W.I. Manton. 1987. "Age of Feiran basement rocks, Sinai: Implications for late Precambrian crustal evolution in the northern Arabian–Nubian Shield." *Journal of the Geological Society, London* 144:569–75.

Strabo. *The Geography of Strabo. Book XVII.* Trans. Horace Leonard Jones, 1932. London: William Heinemann Ltd.

Tawadros, E. Edward. 2001. *Geology of Egypt and Libya.* Rotterdam: A.A. Balkema.

Vivian, Cassandra. 2000. *The Western Desert of Egypt: An Explorer's Handbook.* Cairo: The American University in Cairo Press.

Weeks, Kent. 2000. *KV5: A Preliminary Report on the Excavation of the Tomb of the Sons of Rameses II in the Valley of the Kings.* Cairo: The American University in Cairo Press.

Wendorf, Fred and Romuald Schild. 1976. *Prehistory of the Nile Valley.* New York: Academic Press.

White, Gilbert F. 1988. "The Environmental Effects of the High Dam at Aswan." *Environment* 30(7):5–11, 34–40.

Willcocks, W. 1899. *Egyptian Irrigation.* London: E. & F. Spon, Ltd.

Williams, M.A.J. and Hugues Faure. 1980. *The Sahara and the Nile.* A.A. Balkema, Rotterdam.

Wilkinson, Richard. H., ed. 1995. *Valley of the Sun Kings: New Explorations in the Tombs of the Pharaohs.* Tuscon: The University of Arizona Egyptian Expedition.

————. 2000. *The Complete Temples of Ancient Egypt.* New York: Thames and Hudson.

Wilson, J. Tuzo. 1971. "Introductions." *Continents Adrift: Readings from Scientific American.* San Francisco: W. H. Freeman and Co.

Winkler, E.M. 1994. *Stone in Architecture: Properties, Durability.* 3rd Edition. Berlin: Springer-Verlag.

Zitterkopf, Ronald E. and Steven E. Sidebotham. 1989. "Stations and Towers on the Quseir–Nile Road." *Journal of Egyptian Archaeology* 75:155–89.

No author. 1952. *The Suez Canal: Notes and Statistics.* London: Percy Lund, Humphries, and Co. Ltd.

Index

224

Index